AAPS Advances in the Pharmaceutical Sciences Series

Volume 46

Series Editor
Yvonne Perrie, Strathclyde Institute of Pharmacy and Biomedical Sciences, University of Strathclyde, Glasgow, UK

The AAPS Advances in the Pharmaceutical Sciences Series, published in partnership with the American Association of Pharmaceutical Scientists, is designed to deliver volumes authored by opinion leaders and authorities from around the globe, addressing innovations in drug research and development, and best practice for scientists and industry professionals in the pharma and biotech industries.

More information about this series at http://www.springer.com/series/8825

Saeed Ahmad Khan

Editor

Essentials of Industrial Pharmacy

 Springer

Editor
Saeed Ahmad Khan
Department of Pharmacy
Kohat University of Science and Technology
Kohat, Pakistan

Editorial
Contact Carolyn Spence

Copyright held by the American Association of Pharmaceutical Scientists
ISSN 2210-7371   ISSN 2210-738X   (electronic)
AAPS Advances in the Pharmaceutical Sciences Series
ISBN 978-3-030-84979-5   ISBN 978-3-030-84977-1   (eBook)
https://doi.org/10.1007/978-3-030-84977-1

This Springer imprint is published by the registered company Springer Nature Switzerland AG
The registered company address is: Gewerbestrasse 11, 6330 Cham, Switzerland

This book is dedicated
To my parents who prioritized my happiness
and wellbeing over theirs, and taught me to
be an independent and determined person
To my kids, Rameen, Saqib, Muneeb, and
Rafia, whose love flow like butterflies
around me
To my wife who supported me through thick
and thin
To my siblings for their everlasting love and
warm encouragement throughout my life
To my teachers and students without whom
this book wouldn't have been possible

Preface

Industrial pharmacy is a discipline that deals with the designing, development, manufacturing, marketing, and distribution of drug products including quality assurance and regulatory affairs thereof. Development of drug into a suitable and stable dosage form is a multifaceted effort of many departments in pharmaceutical industry, which requires highly skilled persons in pharmaceutical manufacturing operations and quality management. *Essentials of Industrial Pharmacy* is written with the objective to provide simple and easy illustrations of complex pharmaceutical concepts, processes, and equipments used in pharmaceutical manufacturing. Pictorial/graphical and tabular presentations are attempted to comprehensively present the available information about pharmaceutical processes, formulation, and quality control of different dosage forms.

Essentials of Industrial Pharmacy is divided into 16 chapters. The first five chapters provide concise explanation about basic principles of pharmaceutical processes, design features, and operation principles of the equipments used in pharmaceutical unit operations, viz., drying, mixing, comminution, filtration, and packaging. Chapters 6, 7, 8, 9, 10 and 11 explain formulation aspect and manufacturing of common dosage forms, viz., tablets, capsules, emulsions, suspensions, sterile products, topically applied products, and oral inhalers. Moreover, hazards commonly posed by industrial processes are detailed in Chapter 12. Furthermore, The modified drug delivery systems, novel drug delivery systems, and the polymers used in these delivery systems are covered in Chapters 13, 14 and 15.

Unique features of this book are:

- Comprehensive and comprehensible resource for understanding pharmaceutical processes and manufacturing technology employed for common dosage forms
- Pictorial/graphical and tabular presentation provides simple and easy self-learning format
- Examples from pharmaceutical industry highlights ultimate applications of key theoretical concepts

I hope *Essentials of Industrial Pharmacy* is an essential resource for students studying pharmacy and other disciplines of science (with no pharmacy background) intending to work in the pharmaceutical industry.

Kohat, Pakistan

Saeed Ahmad Khan

Contents

1 **Drying**.. 1
Majeedullah and Asad Shareef

2 **Mixing** 15
Majeedullah and Gul-e-Rana Khalid

3 **Comminution**.................................. 27
Faiza Hanif and Majeedullah

4 **Filtration** 45
Farmanullah and Sajid Hussain

5 **Pharmaceutical Packaging**..................... 55
Fawad Ali and Inzemam Khan

6 **Oral Emulsions** 67
Saeed Ahmad Khan and Saddam Uddin

7 **Oral Suspensions**............................. 81
Saeed Ahmad Khan, Abdul Baseer, Salar Khan,
and Mehmood Hussain

8 **Tablets and Capsules** 95
Amjad Khan, Majeedullah, and Saeed Ahmad Khan

9 **Oral Inhalers**................................ 123
Elena Haettig and Marc Schneider

10 **Topically Applied Products** 151
Sajid Khan Sadozai, Arsh Zafar, and Sheheryar Sajjad

11 **Sterile Products**............................. 177
Muhammad Naseer Abbas, Waqar Iqbal, and Shahzeb Khan

12 **Hazards in Pharmaceutical Industry** 203
Inzemam Khan and Syed Majid Shah

13 **Modified-Release Drug Delivery Systems**. 217
    Saeed Ahmad Khan, Roohullah, and Alam Zeb

14 **Novel Drug Delivery Systems**. 235
    Saeed Ahmad Khan and Hussain Ali

15 **Polymer for Biomedical Applications** 251
    Iqra Riasat, Muhammad Naeem, Muhammad Umar Aslam Khan,
    Syed Babar Jamal, Atif Ali Khan Khalil, Sajjad Haider,
    and Adnan Haider

16 **Biotechnology-Based Therapies**. 261
    Behzad Qureshi, Saadullah Khan, Zia ur Rehman,
    and Noor Muhammad

Index. 285

Contributors

Muhammad Naseer Abbas Department of Pharmacy, Kohat University of Science and Technology, Kohat, Pakistan

Fawad Ali Department of Pharmacy, Kohat University of Science and Technology, Kohat, Pakistan

Hussain Ali Department of Pharmacy, Quaid-I-Azam University Islamabad, Pakistan, Islamabad

Abdul Baseer Department of Pharmacy, Abasyn University Peshawar, Pakistan

Farmanullah Department of Pharmacy, Kohat University of Science and Technology, Kohat, Pakistan

Elena Haettig Biopharmaceutics and Pharmaceutical Technology, Saarland University, Saarbrücken, Germany

Adnan Haider Department of Biological Sciences, National University of Medical Sciences, Rawalpindi, Pakistan

Sajjad Haider Department of Chemical Engineering, King Saud University of Science and Technology, Riyadh, Saudi Arabia

Faiza Hanif Department of Pharmacy, Kohat University of Science and Technology, Kohat, Pakistan

Mehmood Hussain Department of Pharmacy, Kohat University of Science and Technology, Kohat, Pakistan

Sajid Hussain Department of Pharmacy, Kohat University of Science and Technology, Kohat, Pakistan

Waqar Iqbal Department of Pharmacy, Kohat University of Science and Technology, Kohat, Pakistan

Syed Babar Jamal Department of Biological Sciences, National University of Medical Sciences, Rawalpindi, Pakistan

Atif Ali Khan Khalil Department of Biological Sciences, National University of Medical Sciences, Rawalpindi, Pakistan

Gul-e-Rana Khalid Department of Pharmacy, Kohat University of Science and Technology, Kohat, Pakistan

Amjad Khan Department of Pharmacy, Kohat University of Science and Technology, Kohat, Pakistan

Inzemam Khan University of Peshawar, Peshawar, Pakistan

Muhammad Umar Aslam Khan Department of Polymer Engineering and Technology, University of the Punjab, Lahore, Pakistan

Saadullah Khan Department of Biotechnology and Genetic Engineering, Kohat University of Science and Technology, Kohat, Pakistan

Saeed Ahmad Khan Department of Pharmacy, Kohat University of Science and Technology, Kohat, Pakistan

Salar Khan Department of Pharmacy, Abdul Wali Khan University, Mardan, Pakistan

Shahzeb Khan University of Malakand, Chakdara, Pakistan

Majeedullah Department of Pharmacy, Kohat University of Science and Technology, Kohat, Pakistan

Noor Muhammad Department of Biotechnology and Genetic Engineering, Kohat University of Science and Technology, Kohat, Pakistan

Behzad Qureshi Department of Biotechnology and Genetic Engineering, Kohat University of Science and Technology, Kohat, Pakistan

Muhammad Naeem Department of Biological Sciences, National University of Medical Sciences, Rawalpindi, Pakistan

Zia Ur Rehman Department of Pharmacy, Kohat University of Science and Technology, Kohat, Pakistan

Iqra Riasat Department of Bioinformatics and Biosciences, Capital University of Science and Technology, Islamabad, Pakistan

Roohullah Department of Pharmacy, Abasyn University, Peshawar, Pakistan

Sajid Khan Sadozai Department of Pharmacy, Kohat University of Science and Technology, Kohat, Pakistan

Sheheryar Sajjad Department of Pharmacy, Kohat University of Science and Technology, Kohat, Pakistan

Marc Schneider Biopharmaceutics and Pharmaceutical Technology, Saarland University, Saarbrücken, Germany

Syed Majid Shah Department of Pharmacy, Kohat University of Science and Technology, Kohat, Pakistan

Asad Shareef Department of Pharmacy, Kohat University of Science and Technology, Kohat, Pakistan

Saddam Uddin Department of Pharmacy, Kohat University of Science and Technology, Kohat, Pakistan

Arsh Zafar Department of Pharmacy, Kohat University of Science and Technology, Kohat, Pakistan

Alam Zeb Riphah Institute of Pharmaceutical Sciences, Riphah International University, Islamabad, Pakistan

About the Editor

Dr. Saeed Ahmad Khan is an associate professor and chair of the Department of Pharmacy, Kohat University of Science and Technology, Pakistan, and previously worked at Sarhad University of Science and Information Technology, Pakistan. Dr. Saeed Khan graduated in pharmacy from the University of Peshawar, Pakistan. He subsequently joined Abbott Laboratories (Pakistan) Limited, where he worked in solid and liquid dosage form manufacturing sections. He also worked at Parke-Davis & Co. Ltd., Karachi, Pakistan, for pharmaceutical production and quality management. He earned his MS in chemical engineering from Daegu University, South Korea, and obtained a DAAD scholarship to pursue a PhD under the supervision of Prof. Dr. Marc Schneider in the Department of Pharmaceutical Technology, Philipps University Marburg, Germany, where he excelled in the field of pharmaceutical nanotechnology.

He was awarded the prestigious FULBRIGHT fellowship for 1-year postdoc training under the supervision of Professor Robert O. Williams (Johnson & Johnson Centennial Chair in Pharmacy; Editor-in-Chief, AAPS PharmSciTech; and Division Head) in the Department of Molecular Pharmaceutics and Drug Delivery, University of Texas at Austin, USA. He has published many peer-reviewed articles in well-reputed journals, and he has patents and has presented sessions at international conferences.

Chapter 1
Drying

Majeedullah and Asad Shareef

Abstract Drying is one of the most common unit operations in pharmaceutical manufacturing. Selection of dryers for pharmaceutical manufacturing is not simple, since different existing dryers can dry the material to the desired level. Selection of efficient dryer and drying parameters requires basic knowledge about drying process. This chapter explains the basic concepts of drying, including its importance and application in pharmaceutical processes. The theory of drying and the parameters affecting the drying process are discussed in detail. It also briefly outlines the several types of dryers used in pharmaceutical industries.

Keywords Theory of drying · Drying curve · Types of dryers · Dryer selection · Classification of dryers

1.1 Introduction

Drying is normally referred to the process of thermally eliminating moisture to obtain solid product. Drying is typically intended for adjustment of moisture levels in solid materials. In practice this liquid refers to the water, but at times certain volatile substances may need to be removed [1]. Drying and vaporization can be distinguished by the relative moisture content removed from the solid.

Drying is one of the most important steps in primary and secondary pharmaceutical processes:

- For recovery of drug during synthesis
- For adjustment of moisture content in powder and granules

Majeedullah (✉) · A. Shareef
Department of Pharmacy, Kohat University of Science and Technology, Kohat, Pakistan
e-mail: drmajeed@kust.edu.pk

© The Author(s), under exclusive license to Springer Nature Switzerland AG 2022
S. A. Khan (ed.), *Essentials of Industrial Pharmacy*, AAPS Advances in the
Pharmaceutical Sciences Series 46, https://doi.org/10.1007/978-3-030-84977-1_1

- Impart special properties to material, e.g., spray-dried lactose for better flow
- Reduce weight of material
- Improve stability by reducing hydrolysis of drug
- Improve shelf life by reducing microbial growth

Drying can be attained by many ways: using adsorbent, such as calcium chloride to adsorb moisture, using desiccant in a sealed container such as silica gel that remove water from the air, or exposing material to heat. Heating is the most common mode of drying in pharmaceutical processes.

Moisture held in the microstructure of solid that cannot be easily removed called *Bound water*, has a vapor pressure lower than that of pure water. Moisture other than the bound water is called *Unbound water* [2]. Removal of bound water requires more energy as compared to the unbound water.

When a moist material is heated, two phenomena take place simultaneously:

1. Transfer of energy to the material from surrounding to vaporize the surface moisture
2. Transfer of internal moisture to the surface and ultimately to the surrounding

Therefore, the rate of drying is controlled by the rate at which both these steps take place. The heat transfer from the surrounding to the material can occur as a result of *convection, conduction,* or *radiation,* and in many cases combination of all these.

For understanding it can be simplistically represented by combining heat transfer and mass transfer equations [3]:

$$\frac{dm}{dt} = \frac{qc + qk + qr}{Lv} = K'A\Delta C$$

(1.1)

where dm/dt is the rate of vapors transferred per unit time; qc, qk and qr are the rate of heat transfer by *convection, conduction,* and *radiation,* respectively; Lv is the latent heat of vaporization; K' is mass transfer coefficient; and ΔC is the difference in moisture content between material and environment.

It is evident from the above equation that rate of evaporation can be increased by:

- Increasing q_c = by rising the flow rate of air (i.e., *convection*)
- Increasing q_k = by prolonging contact time with the heating surface (i.e., *conduction*)
- Increasing q_r = by introducing high temperature radiating heat source, such as heating coils (i.e., *radiation*)
- Increasing A = by increasing the surface area
- Increasing ΔC = by dehumidifying the inlet air

1.2   Important Terms Used in Drying Process

Proper understanding of drying process requires the explanation of following related terms.

Total Moisture Content   It is the total amount of liquid present in the material. Not all the moisture can be removed by simple evaporation [4]. The amount of easily removable moisture (unbound) is known as free moisture content, and the remaining moisture is termed as equilibrium moisture content.

Equilibrium Moisture Content   It is the amount of moisture present in a material at ambient conditions, and it is dependent on the humidity, temperature, and nature of material.

Bound Water   It is the moisture trapped in the microstructure of the solid that cannot be easily removed. Bound water has vapor pressure lower than that of pure water.

Percent Loss on Drying (% LOD)   It is the term used to express moisture in a wet substance on the basis of wet weight [5]. It is calculated as:

$$\%LOD = \frac{\text{Weight of water in a solid}}{\text{Total weight of the wet solid}} \times 100 \tag{1.2}$$

The LOD of the wet solid can be determined using a balance which has calibrated scale. A weighed sample is placed in the pan and dried until constant weight. The moisture lost by evaporation is read directly from the % LOD scale.

Percent Moisture Content (% MC)   It is the term used to express moisture content of a substance on dry-weight basis. It can be calculated as:

$$\%MC = \frac{\text{Weight of water in a sample}}{\text{Weight of the dry sample}} \times 100 \tag{1.3}$$

For example, if 8 g of moist solid is brought to a constant dry weight of 5 g then:

$$\%MC = \frac{8-5}{5} \times 100 = 60\% \tag{1.4}$$

while,

$$\%LOD = \frac{8-5}{8} \times 100 = 37.5\% \tag{1.5}$$

LOD values may vary from 0% to slightly below 100%, but MC values may vary from above 0% to infinity. A small change in LOD value represents an increase in

MC value. Thus %MC is more realistic value than %LOD in evaluation of drying process.

Dry-Bulb Temperature It is simply the ambient condition, i.e., the temperature of dry air, and is typically measured with a conventional thermometer.

Wet-Bulb temperature It gets its name because a wet permeable membrane, such as wet gauze, is used in conjunction with a regular thermometer for temperature measurement [6]. The air temperature is measured while a wet gauze is wrapped around the thermometer bulb. Since water is evaporated from the gauze, it leaves evaporative cooling on the thermometer. Consequently wet-bulb temperature is lower than dry-bulb temperature [7]. However, at 100% relative humidity dry-bulb temperature is equal to wet-bulb temperature, since no evaporation of water takes place from wet gauze.

1.3 The Drying Curve

Each material has a representative *drying curve* that describes its drying behavior. This curve is obtained by plotting moisture content vs time. The changes during drying process can also be understood by plotting drying rate vs moisture content [8]. As shown in Fig. 1.1, the drying curve can be divided into four stages:

(a) Initial adjustment period
(b) Constant rate period
(c) First falling rate period
(d) Second falling rate period

The stage A–B in Fig. 1.1 represents the initial adjustment period. The drying surface is equilibrated with drying air. This stage is very short and often ignored in predicting the drying time. The stage B–C shows a constant rate period. A moisture film continuously exists on the solid surface, and solvent is evaporated continuously

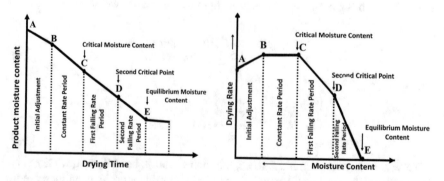

Fig. 1.1 Typical drying curves

from this moisture film. The moisture content at the end of constant rate period (point C) is called critical moisture content (CMC). From this point onward the first falling rate period starts, i.e., the rate of drying slows down, since the moisture film on the solid surface starts to vanish. The second falling rate period begins from point D, where the surface is completely dried. The moisture diffuses from within the material to the surface and gets evaporated. Most of the drying occurs in second falling rate period [9].

1.4 End Point Detection in Drying Process

The drying end point is one of the key factors that affect quality of final product. For batch dryers, end point is typically measured by monitoring moisture content periodically. Then constructing a *"drying curve"* by plotting residual moisture content in the product over time. The end point is the point where the desired level of residual moisture in the product is attained.

To minimize sampling, inline end point testing is preferred, which relies on secondary factors, such as airflow, air pressure, inlet or outlet air temperature, and product temperature. Recently, the inline near infrared spectroscopy (NIR) for determination of product moisture is more common. Moreover, inline mass spectrometry (MS) has also been used for continuous monitoring of headspace solvent concentration.

1.5 Selection of Industrial Dryers

Selection of dryers for pharmaceutical manufacturing is not simple, since different existing dryers can dry the material to the desired level. The choice of dryer depends on the nature of the material being handled and the cost to attain required end point while complying all safety and environmental standards. A more structural approach in selection of dryer may follow the following steps [10].

(a) Process specifications are selected.
(b) A few of the dryers in the available options are selected.
(c) Bench-scale operation and quality tests are performed.
(d) Economic feasibility of the selected dryer is evaluated.
(e) Pilot-scale trials are performed.
(f) The most appropriate dryer is selected.

However, most recently, in many advance setups, dryers are selected on the basis of algorithms that assign a certain score to each dryer, based on the information available for different dryers, the properties of material to be dried, the quantity to be dried, and the required moisture content, etc. [11].

1.6 Classification of Dryers

1.6.1 Classification on the Basis of Mode of Energy Input

Certain features of dryers might vary with the mode of energy input that may affect the feasibility of dryer for certain applications. The mode of energy input may be direct or indirect; hence the corresponding dryers are called direct dryers or indirect dryers.

1.6.1.1 Direct Dryers

These are the most common type of dryers often referred to as *convective dryers*. Most of the industrial dryers are of this type. As the name indicates, the drying medium directly contacts the content to be dried. Typically, hot air or gas is used as drying medium that provides heat for drying and carries the evaporated moisture away. The temperatures of air/gas may vary from 50 to 400 °C. Sometimes, dehumidified air/gas might be required for drying of very heat-sensitive contents. An inert gas like nitrogen may possibly be suitable while drying explosive or flammable materials. Solvents recovered must be extracted by condensation from the exhaust so that they can be reused.

1.6.1.2 Indirect Dryers

In these dryers the drying medium does not directly contact the material to be dried, i.e., heat is transferred by *conduction*. However, it is essential to either use vacuum/ or gentle air current to get rid of the vaporized moisture to avoid chamber saturation with vapors. Heat transfer shelves could range in temperature from −40 °C (as in freeze-dryer) to nearly 300 °C (combustion of waste sludges). Application of vacuum is also helpful for the retrieval of solvents through direct condensation. Additionally, vacuum setup drops the boiling point of the liquid withdrawn and thus permits drying of heat-sensitive materials at fairly faster rates. These types of dryers are applicable for drying of toxic, dusty products. In addition, heat may also be delivered by radiation (electric or natural gas-fired radiators) or dielectric fields in the microwave/or radio-frequency range.

1.6.2 Classification on the Basis of Solid Handling

Dryers can also be classified on the basis of how the solid is handled during drying, for instance:

(a) Static bed dryers
(b) Moving bed dryers
(c) Fluidized bed dryers
(d) Pneumatic dryers
(e) Specialized dryers

1.6.2.1 Static Bed Dryers

In static bed dryers the individual solid particles are kept static; however, the entire bed of drying powder is moved. Consequently, only a fraction of surface material is exposed to the heat. Reduction in thickness of powder bed will result in increased exposure of the surface. Examples are tray dryer and truck dryer.

Tray Dryer

Tray dryer also called shelf dryer is one of the most commonly used dryers for small-scale drying. It comprises a central cabinet for placement of trays, as shown in Fig. 1.2a. The number of trays held in cabinet is based on the size and shape of the dryer. The heat supply is either direct or indirect. In direct heat supply, hot air is distributed inside the chamber. Indirect heating is provided either by heated shelves or by radiant heat source inside the chamber. The base of trays may be solid or perforated. The trays with solid base utilize heat from top and bottom of the tray, while in perforated base trays, air passes through each tray and the solid. The perforated trays need to be lined with papers as a disposable tray liner to reduce cleaning time and prevent product contamination. For uniform drying a well-insulated cabinet with placed fans and heating coils inside the cabinet is preferred.

Truck Dryer

It is very similar to tray dryer; however, the trays are loaded on truck (wheeled racks), which can roll into and out of the drying cabinet, as shown in Fig. 1.2b. This offers more convenience in loading and unloading of trays. A wheeled rack can house around 18 trays, with about 4–8 ft^2 placement area. The trays are generally loaded 0.5–4.0 inches deep with at least 1.5 inches between the surface and above trays. A control panel is fixed outdoor to monitor the temperature and other features.

Tunnel Dryer

It is an adaptation of the truck dryer and is suitable for large-scale production. It consists of a long tunnel (drying chamber), as shown in Fig. 1.2c. The solid is placed on the trays that are loaded on truck [3]. Air current enters from one end, and trucks

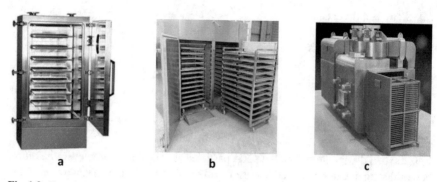

Fig. 1.2 Tray dryer (**a**), truck dryer (**b**), and tunnel dryer (**c**). (Courtesy: Bigtem machines, Turkey)

are slowly moved from the other end with the help of moving chain. The trucks reside in the drying chamber for sufficient time to achieve adequate drying. Tunnel dryers are classified as semicontinuous dryers, since they involve manual handling of material in the drying cycle. However, the drying achieved is not uniform, as the top exposed layer is dried fast compared to the lower layer.

Conveyor Dryer

It is a continuous dryer and is adaptation of tunnel dryer, where individual trucks are replaced with a conveyor belt or panel that carries the wet mass via the tunnel, as shown in Fig. 1.3. In some conveyor design, the partially dried mass rolls onto adjacent conveyor heading in the opposite direction. The wet mass can have a number of serial passages along the chamber before it is discharged from the conveyor.

1.6.2.2 Moving-Bed Dryers

In moving-bed dryers, the particles flow over each other by gravity or due to mechanical agitation. Hence, new surface is regularly exposed to heat that facilitates rapid heat transfer to the particles and improves evaporation of moisture from the particles. Examples of moving-bed dryers are given in coming section.

Rotary Dryer

It consists of a slightly inclined horizontal tube with a feed inlet at one end. The material enters at feed inlet and passes in the revolving cylinder, as shown in Fig. 1.4. The rotation and slope of the revolving cylinder are responsible for the material to slide down [12]. Heating is directly provided using hot air or indirectly by an external jacketed steam tube. Rotary dryers provide continuous and faster

Fig. 1.3 Conveyor dryer. (Courtesy: Bigtem machines, Turkey)

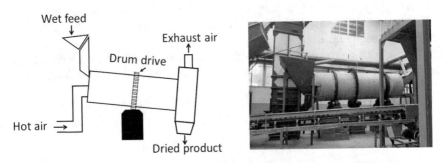

Fig. 1.4 Rotary drum dryer. (Courtesy: Changzhou Fuyi Equipment Co., Ltd. China)

drying. Moreover, the drying achieved is uniform, since all parts get similar exposure to hot air.

1.6.2.3 Fluidized Bed Dryer

In fluidized bed dryer (FBD), individual particles are lifted and then fall back in a random manner, called *fluidization* of particles. FBD generally offers better heat and mass transfer than static and moving-bed dryer. There are two subtypes of FBD, vertical FBD and horizontal FBD.

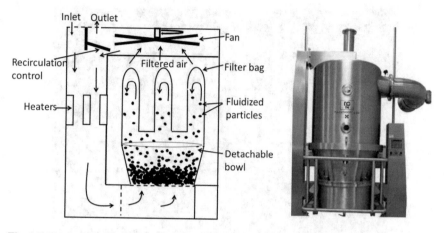

Fig. 1.5 Vertical fluidized bed dryer. (Courtesy: PA Cuthbert & Co. (Pvt) Ltd, South Africa)

Vertical FBD

In vertical FBD, hot air is blown through the wet mass held in drying chamber having a wire mesh support at the bottom. A bag collector filter is mounted at the top of the drying chamber to stop particles from being carried over [13]. A typical vertical FBD is given in Fig. 1.5.

It is a batch-type dryer and typically has the capacity of 5–200 kg, with average drying time in between 20 to 40 min. It offers efficient, fast, and uniform drying with reduced labor cost [3].

Horizontal FBD

Horizontal FBD comprises a vibrating conveying deck passing over a perforated surface; hot air is blown up through perforated conveying surface. The vibration of conveying deck fluidizes the particles. Air stream passes over the fluidized bed into the exhaust hood, as shown in Fig. 1.6.

Horizontal FBD offers continuous drying; however it is not suitable for friable materials.

1.6.2.4 Pneumatic Systems

In pneumatic systems, particles slurry is introduced to the drying chamber in the form of a fine mist. Individual particles are entirely exposed to drying gas as a result heat and mass transfer is very rapid. One of the most common examples of pneumatic systems is spray dryer.

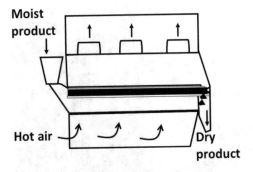

Fig. 1.6 Horizontal fluidized bed dryer. (Courtesy: Tianjin HSD Separation Envitech Co., Ltd, China)

Spray Dryer

Contrary to previously discussed dryers, spray dryer can handle only fluids, such as solutions, slurries, and thin pastes. It comprises of a drying chamber, an atomizer and a hot air source and exhaust. The fluid is converted into a fine mist in a moving stream of hot air, where liquid from the droplet is evaporated before touching the wall of drying chamber. Fine dried particles are moved by the gas and gravity flow into the main collector by a cyclone separator (cyclone product), as shown in Fig. 1.7. Some of the dried powder remains in drying chamber, called "chamber product," and is collected from the bottom of the chamber. Particles in chamber product are usually coarser and are subjected to heat for longer time. Final product is a mixture of cyclone product and chamber products [14].

The spray dryer atomizer is of three types:

Pneumatic atomizers: Droplets are formed by a high-velocity jet of air or gas.
Pressure nozzle: Liquid feed is supplied by high-pressure nozzles (upto7000 lb/In²) and splits by impact of a fixed plate placed in front of the stream.
Spinning disc atomizers: Liquid is split into droplets by a rapidly rotating disc (3000–50,000 rpm).

Spray drying offers three major advantages;

Drying heat-sensitive material. It is suitable for drying of thermolabile materials without degrading them, as evaporation of the liquid keeps the temperature of material lower.
Changing the physical form of material. It changes the shape, size, and bulk density of dried product. It promotes flowability as same size and shape particles are formed with rare sharp edges. The process also reduces air trap in spherical particles making them preferable for usage in the solid dosage forms.
Coating and encapsulation of both solids and liquids. Particles suspended in a solution of the coating agent can be spray-dried into coated particles. As the solvent is evaporated, the coating material covers the suspended particles.

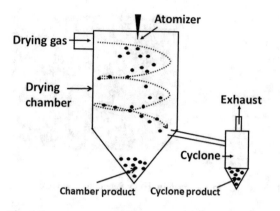

Fig. 1.7 Spray dryer. (JY Drying Engineering Co., Ltd, China)

1.6.2.5 Specialized Dryers

Specialized dryers are designed to dry sensitive pharmaceutical and chemicals, e.g., vacuum dryer and freeze dryer.

Freeze Dryer

Freeze dryer uses the principle of *sublimation*, the conversion of a substance from solid state directly to gas state, without converting into liquid state, as shown in Fig. 1.8.

The liquid contained in material is frozen and evaporated at low temperature under high vacuum. This makes freeze dryer suitable for thermolabile products.

The boiling point of liquid is governed by temperature-pressure relationship. For instance, at normal atmospheric pressure (1 atm or 760 mmHg), water evaporates at 100 °C. The boiling point can be reduced by reducing the pressure and vice versa. At *triple point* (i.e., temperature = 0.0098 °C and pressure = 4.58 mmHg), pure water starts exhibiting the process of sublimation [15]. However, in pharmaceutical dosage forms, certain dissolved components may affect temperature-pressure relationship for each solution. The point at which these solutions are sublimed is called *eutectic point* [16].

Freeze drying is typically carried out at temperature range of −10 to −40 °C under 2.0 to 0.1 mmHg pressure [17].

Advantages of freeze drying

- Freeze drying provides minimum damage to heat-labile materials.
- In freeze drying, minimal structural changes or shrinkage occurs.
- Drying takes place at very low temperature, so that enzyme action is inhibited.
- The dried powder is fluffy and fibrous and has better solubility.
- Freeze drying may increase the shelf life of some drugs.

Fig. 1.8 Schematic pressure-temperature diagram for water

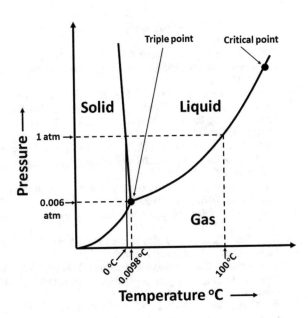

- Sterility can be maintained during drying process.

Disadvantages of freeze drying

- Freeze drying system needs vacuum and refrigeration equipment. The initial costs are four times higher than conventional drying. Moreover, energy costs are also higher than the other methods.
- Drying rate is very slow.
- Packaging of freeze-dried product requires special attention/conditions.

Uses of freeze drying

- Freeze drying is used in preparation of hormones, blood products, antibiotics, bacterial culture, and vaccines.
- It is used for food products such as freeze-dried ice cream and instant coffee.
- Recently, some taxidermies have begun using freeze drying to preserve animals.

References

1. Kemp IC, Oakley DE. Modelling of particulate drying in theory and practice. Drying Technology. 2002;20(9):1699–750.
2. Khan MIH, Wellard RM, Nagy SA, Joardder M, Karim M. Experimental investigation of bound and free water transport process during drying of hygroscopic food material. International Journal of Thermal Sciences. 2017;117:266–73.

3. Lachman L, Lieberman HA, Kanig JL. The theory and practice of industrial pharmacy: Lea & Febiger; 1986.

4. Hiew TN, Johan NAB, Desai PM, Chua SM, Loh ZH, Heng PWS. Effect of moisture sorption on the performance of crospovidone. International Journal of Pharmaceutics. 2016;514(1):322–31.

5. Razvi SZA, Kamm I, Nguyen T, Pellett JD, Kumar A. Loss on Drying Using Halogen Moisture Analyzer: An Orthogonal Technique for Monitoring Volatile Content for In-Process Control Samples during Pharmaceutical Manufacturing. Organic Process Research & Development. 2021;25(2):300–7.

6. Huang Y, Zhang K, Yang S, Jin Y. A method to measure humidity based on dry-bulb and wet-bulb temperatures. Research Journal of Applied Sciences, Engineering and Technology. 2013;6(16):2984–7.

7. Hasan A. Indirect evaporative cooling of air to a sub-wet bulb temperature. Applied Thermal Engineering. 2010;30(16):2460–8.

8. Kemp IC. Drying of pharmaceuticals in theory and practice. Drying Technology. 2017;35(8):918–24.

9. Mujumdar AS. Drying of Pharmaceutical Products. Handbook of Industrial Drying, Fourth Edition, CRC Press, 2014 pp.711–732.

10. Mujumdar AS. Classification and selection of industrial dryers. Mujumdar's Practical Guide to Industrial Drying: Principles, Equipment and New Developments Brossard, Canada: Exergex Corporation.. 2000:23–36.

11. de Oliveira WP, de Freitas LAP, Freire JT. Drying of pharmaceutical products. Transport Phenomena in Particulate Systems, Freire, JT, Silveira, AM, Ferreira, MC, eds, Bentham Science. 2012:148–71.

12. Abidin ASZ, Kifli MZ, Jamali A, Muslimen R, Ahmadi R. Development of Black Pepper Rotary Drum Dryer System. International Journal of Integrated Engineering. 2020;12(7):11–9.

13. Gagnon F, Bouchard J, Desbiens A, Poulin É. Development and validation of a batch fluidized bed dryer model for pharmaceutical particles. Drying Technology. 2020:1–24.

14. Celik M, Wendel SC. Spray drying and pharmaceutical applications. Handbook of pharmaceutical granulation technology. 2005;2:157–186.

15. Assegehegn G, Brito-de la Fuente E, Franco JM, Gallegos C. Freeze-drying: A relevant unit operation in the manufacture of foods, nutritional products, and pharmaceuticals. Advances in Food and Nutrition Research. 2020;93:1–58.

16. Jones DS. FASTtrack Pharmaceutics dosage form and design: Pharmaceutical Press; 2016.

17. Nowak D, Jakubczyk E. The freeze-drying of foods—The characteristic of the process course and the effect of its parameters on the physical properties of food materials. Foods. 2020;9(10):1488.

Chapter 2
Mixing

Majeedullah and Gul-e-Rana Khalid

Abstract Pharmaceutical processes that need mixing include dissolution of solids in liquids, mixing of liquid with liquid, dispersion of solid or liquid in liquids, blending of solid with solid, and mixing of gas with liquid. This chapter expands upon the concept of mixing and its classification. Different types of mixing along with the mechanism of mixing are discussed in detail. The chapter also reviews the principles and mechanisms of solid and liquid mixing. Moreover, the equipments used in liquid and solid mixing are also discussed in detail.

Keywords Mechanism of mixing · Mixing of liquids · Mixing of solids · Demixing · Scale of scrutiny

2.1 Introduction

Mixing is a unit operation that aims at bringing two or more unmixed or partially mixed components as close to each other as possible [1].

Majority of pharmaceutical dosage forms exist as multicomponent systems containing drug (active ingredient) and excipients (inactive ingredients) that need homogeneous distribution of components in order to ensure reproducible therapeutic response [2]. The dose uniformity, stability, and product overall acceptability are greatly affected by mixing process; therefore, selection of appropriate mixer is a critical step in the manufacturing of quality pharmaceutical products. This chapter

Majeedullah (✉) · Gul-e-Rana Khalid
Department of Pharmacy, Kohat University of Science and Technology, Kohat, Pakistan
e-mail: drmajeed@kust.edu.pk

© The Author(s), under exclusive license to Springer Nature Switzerland AG 2022
S. A. Khan (ed.), *Essentials of Industrial Pharmacy*, AAPS Advances in the
Pharmaceutical Sciences Series 46, https://doi.org/10.1007/978-3-030-84977-1_2

focuses on the basic concepts of mixing process used in pharmaceutical manufacturing.

Mixing is employed in many manufacturing processes, for instance:

- Dissolution of drug in solvent, e.g., syrups and elixirs.
- Mixing of powders for manufacturing of tablets, capsules, sachets, dry powder inhalers, etc.
- Mixing of two immiscible liquids, e.g., emulsion and creams.
- Dispersion of solids in liquids, e.g., suspensions and pastes.

Based on the mutual mixing tendency of components, the product of mixing can be *positive mixture, negative mixture,* and/or *neutral mixture.* The irreversible mixture is formed by spontaneous diffusion of components without application of energy in *positive mixture,* for example, mixture of two miscible liquids, while *negative mixtures* are reversible mixtures that require continuous supply of energy. The components tend to separate as soon as the required energy is stopped, e.g., pharmaceutical suspensions and emulsions. *Neutral mixtures* neither has the tendency of spontaneous mixing nor segregation, e.g., ointments [3].

2.2   Mixer Selection

Efficient mixing requires selection of proper mixer that is able to mix material without affecting product quality. Ideally mixer should be:

- Easily cleanable
- Easily usable
- Energy efficient
- Contamination resistant and dust tight

However, presence of all these features in a single mixer is rare. Several factors need to be considered during selection of an optimum mixer. One of the first things to consider is whether the process intended is a batch or a continuous one, each of which can have its advantages and disadvantages depending on the extint to be mixed [4]. However, mixers for liquid pharmaceuticals are usually batch type, since dosage forms are prepared in individual batches. Another important consideration is optimal working volume, that is, usually between 50% and 70% of the limit, since too much load can lead to low mixing [5].

2.3   Mixing of Liquids

Mixing of liquid may occur due to *molecular diffusion or convection. Molecular diffusion* suggests that molecules spontaneously move from higher concentration to lower concentration. Molecular diffusion is governed by *Fick's law of diffusion* that states that the rate of *molecular diffusion* (dM/dt) is directly proportional to concentration gradient (ΔC), diffusion coefficient of the material (D), and surface area of the material (A) and inversely proportional to thickness of the supersaturated layer on the surface (L).

$$dM / dt = -DA\Delta C / L \qquad (2.1)$$

Mixing may also involve *convection*, i.e., a large portion of material in the system is moved from one location to another with the help of mixing equipment, such as moving blades and impellers. The moving impeller applies shear force, which is applied to push a portion of material in one direction against another part of material in the same parallel plane. High shearing can increase turbulence in some fluids [6]. Turbulent flow is characterized by the presence of large number of *eddies*, which is defined as a portion of fluid that flows in a direction opposite to the general flow of a fluid. However, in some highly viscous liquids, the velocity components at a given point in the flow remain constant during mixing. This type of flow is termed as *laminar flow* [7].

2.3.1 Batch-Type Liquid Mixing

The equipment used for mixing of liquids is usually batch type, since the material to be mixed is restricted to a fixed volume. Batch mixing is typically performed in a vessel supplied with mixing impeller, i.e., a rotational device that provides shear force to the components and produces a homogeneous mixture.

2.3.1.1 Impeller Mixer

The role of mixing impeller is to transfer rotational energy of the shaft into turbulence in order to achieve desired mixing characteristics. The impeller design can affect process performance, which is ultimately governed by the specific characteristics required by a given process. Table 2.1 summarizes different types, shapes, and respective uses of impellers.

The important design features of an impeller-type liquid batch mixer are shown in Fig. 2.1

Number of Impellers affects design features; single impeller is preferable due to cost. However, huge difference in liquid level (L) and vessel diameter (T) can adversely affect flow patterns generated within the vessel and in turn lead to poor mixing efficiency. Hence, multiple impellers would be required in such cases.

Impeller Positioning within the mixing vessel can have a significant effect on the overall process performance; poor position of impellers can lead to poor mixing and impellers may be out of the liquid in critical stages of the process.

Table 2.1 Types, shapes, and features of different types of impellers

Type of impellers	Shape	Key features
Propeller		– Operated at high speed – Blades can be mounted at any angle – Suitable for low viscosity (<2.0 pascals) liquids
Turbine impeller		– Operated at high speed – Mixing efficiency can be improved by mounting a stationary perforated ring around the turbine impeller – Vertexing can be reduced by the diffuser ring – Air bubbles can be trapped within the liquid
Paddle impeller		– A plane blade connected to a spinning shaft – Suitable for miscible liquids and to dissolve soluble solids in liquids
Anchor agitator		– Suitable for viscous liquids and slurries – Operates near to the wall of the container and promotes heat transfer
Helical impeller		– Suitable for mixing of viscous fluids, such as gels

Impeller Diameter (D) to Vessel Diameter (T) Ratio (D/T Ratio) greatly affects the performance of most fluid mixers. Optimum D/T depends on process requirements and is typically in the range of 0.2–0.5.

Impeller Bottom Clearance can affect mixing performance as well as pumping efficiency. The optimum C/T ratio is mainly governed by impeller type, but process condition can also affect this ratio. C/T in the range of 0.1–0.3 is typically considered as optimum value.

Vessel Geometry plays an important role in final design of mixer. Inappropriate vessel geometry reduces mixing efficiency, which in turn increases cost.

Incorporation of Baffles for increasing the turbulence in the flow of fluid, plates called "baffles" are mounted vertically along the walls of vessels to increase resistance to flow. Their placement increases mixing efficiency of mixers.

2.3.1.2 Air Jets

These systems use pressurized air or gas introduced through a jet mounted at the bottom of vessel. A draft tube is mounted in the middle. The liquid entrained in the draft tube is raised by the moving air, as shown in Fig. 2.2. These repeated cycles result in efficient mixing. However, liquid being mixed by air jets must be less

Fig. 2.1 Some of the important aspects in designing a batch mixer for liquid mixing, where D is diameter of impeller, T is diameter of vessel, C is impeller bottom clearance, L is level of liquid, and B is baffles

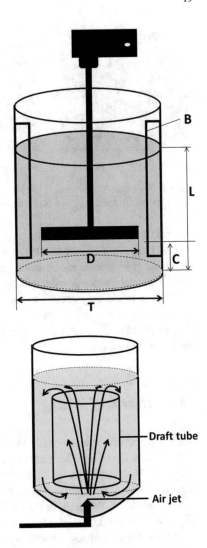

Fig. 2.2 Diagrammatic representation of air jet mixer

viscous, non-reacting with the gas used in the process, and should not form foam during this operation.

2.4 Mixing of Solids

2.4.1 Principle of Solid Mixing

The principle of solid mixing can be well understood by assuming a situation where equal quantities of two different powders having same particle size, shape, and density are represented by colored cubes. Two-dimensional illustration of the preliminary unmixed or totally segregated state can be seen in Fig. 2.3a. From the definition

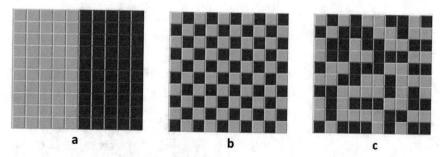

Fig. 2.3 Schematic representation of two-ingredient powders before and after mixing, (**a**) segregated powder, (**b**) ideal mix, and (**c**) random mix

of mixing, the ideal situation or perfect mix would be when each particle of one powder is surrounded by particles of another component. This is shown in Fig. 2.3b, where the components are as evenly distributed as possible. However, it is practically impossible, since powder mixing is a "chance" process and the situation shown in Fig. 2.3b could only occur by chance. Therefore, the common practical situation of solid mixing is random mix, as shown in Fig. 2.3c, which is defined as a mix where the probability of selecting a particular type of particles is the same at all positions in the mix.

2.4.2 Degree of Mixing

Obviously, powder mixture will have a certain degree of heterogeneity. It is important to define the criteria for considering a powder to be "homogeneous" or "heterogenous" [8]. Sample size is critical for determining whether a powder has a random mix. To better understand, let's assume that each square in Fig. 2.3 represents a single particle. If the sample size is 100 particles, then both segregated powder (Fig. 2.3a) and random mix (Fig. 2.3c) would have no difference and both would be considered homogeneous. Obviously, a sample size bellow 25 particles would be needed to differentiate both the samples. Sample size is determined by the *scale of scrutiny*, which is the minimum sample size that must be evaluated to determine if the mixture meets the desired degree of mixing [3]. Scale of scrutiny is critical for evaluating a good mix. As a rule of thumb, the smaller the scale of scrutiny, the better is the homogeneous mixture.

2.4.3 Segregation of Powder (Demixing)

A mixture containing different sizes of particles will always tend to separate. Separation of particles of one size in one area and another size in another area is called *segregation* or *demixing* [9]. It is obvious that pharmaceutical powders always have particles of different sizes. This means that all powders would be subjected to demixing. As a rule of thumb, if the ratio of particle sizes increases than 1.3, the powder will exhibit segregation problem [10].

Segregation may lead to content variation that will ultimately lead to failure of content uniformity test. If segregation of granule occurs during compression process, it may result into weight variation in tablets and capsules. Segregation is more likely to take place when powder is exposed to vibration or tapping.

Segregation is a function of particle size, particle geometry, and particle density. However, the primary factor is particle size and specially the difference in size of particles in a powder.

2.4.4 Mechanism of Solid Mixing

Solid mixing process involves combination of mixing principles, *diffusive mixing*, *shear mixing*, and *convective mixing*.

Diffusive Mixing occurs when particles tend to occupy the voids present in powder bed, through the effect of gravity. This type of mixing usually occurs when particles are rolled down on inclined surface.

Convective Mixing occurs when a bulk of particles from one part of the powder bed is moved to another part, usually with the help of mixing tool. However, particles that are moved as a unit to attain a state of random mix require prolonged mixing time. Since, mixing does not occur in the group of particles that are moved together as a unit, prolonged mixing time is required to get a random mix.

Shear Mixing is sometimes considered as part of the convection mechanism, since particles are moved by a mixing tool, i.e., the shearing force establishes a slipping zone in powder. That causes *shearing* of one layer over another. The shearing force is particularly useful if the powder contains lumps that need to be broken down in order to attain a state of random mix. Such type of mixing mechanism may occur in high shear mixers or tumbling mixers, where the shearing force provided by mixing tool induces velocity gradient within the powder bed and the powder slip in between each other.

All the discussed three mixing mechanisms are likely to occur in a typical mixing process.

2.4.5 Factors Affecting Solid Mixing

Mixing of powder can be affected by many factors, such as particle shape, cohesion of particles, mixer volume, mixing time, and mixer speed.

Particle Shape The spherical the shape of particle, the easier will be the mixing of material.

Cohesion of Particles The particles have the tendency to bind to each other, and they are hard to be mixed.

Mixer Volume The larger the mixer capacity, the easier it is to mix the material.

Mixing Time The longer the mixing time, the better will be the mixing.

Mixer Speed The higher the mixing speed, the effective will be mixing; however, sometimes extremely high speed of mixing inhibits mixing. Moreover, it may also change particle size of the powder.

2.4.6 Equipments for Solid Mixing

For solid-solid mixing, both manual and mechanical mixing are done. Manual mixing is usually done on small scale, while mechanical mixing is for large scale.

2.4.6.1 Manual Mixing

Trituration in a glass mortar and pestle is a common method for laboratory-scale mixing. *Geometric dilution* method is used when small quantity of drug has to be mixed with large quantity of diluent. The drug is first mixed with equal amount of diluent, and the powder is triturated until completely mixed. An equal amount of diluent is again added and triturated. The process is repeated until all the powder is mixed. For powders that resist mixing by trituration, *sifting* might be a useful method. This is done by shaking powder components through a sieve. Sifting is very effective for mixing of light powders, e.g., magnesium oxide and charcoal.

Powder can also be mixed by *tumbling*, i.e., shaking in a closed container or ziplocked bags. This method is particularly useful for powders having significant variations in particle densities. This method does not reduce particle size.

Another effective method for mixing of small quantities of powders is *spatulation*. The powders are blended on a pill tile (ointment slab) using spatula. The potential loss during transfer is small.

2.4.6.2 Mechanical Mixing

The preferred industrial method for solid mixing is by use of rotating elements that randomly transfer powder mass from one place in the mixer to another. Some of the solid mixers are described here.

Fig. 2.4 Sigma blade mixer (**a**), ribbon mixer (**b**). (Courtesy: Yoko Food Engineering, Malaysia)

Horizontal Ribbon Mixer and Sigma Blade Mixer

These mixers consist of horizontal semicylindrical troughs, containing one or two rotating elements attached to central shaft attached to motor. The rotation of mixing elements causes the material to mix. The mixing elements may be in the form of sigma blades or ribbons (Fig. 2.4). The shape of blades/ribbons is designed such that one blade/ribbon tends to move the powder in one direction while the other one moves it in opposite direction. Convection is the main mechanism of these types of mixers. The mixing vessels may be jacketed for supply of steam, hot water, or cold water.

Tumbling Mixers

It consists of multi-geometric container mounted on a central shaft. Rotation of the shaft rotates the whole mixer on its axis. The most common shapes used for tumbling mixers are V shape, cone shape, cube shape, etc., as shown in Fig. 2.5. The shape of container favors mixing when the container is rotated. Inclusion of baffles or bar having rotating blades can improve agitation of the powder bed [11]. The efficiency of mixing is greatly affected by the speed of rotation. Too fast rotation can cause the powder to stick to the container walls, while too slow rotations cannot produce the desired tumbling action. The optimum speed is typically in the range of 30–100 rpm; however, it varies with the shape and size of the mixer. The predominant mechanism of mixing is diffusion and convention [12].

High Shear Mixer Granulator

As the name indicates, it can be used for mixing as well as granulation of powder, thus eliminating many steps of material transferring. It consists of impellers which are centrally mounted and rotates at high speed. The impeller throws the material to the mixing bowl by centrifugal force. The particular movement tends to quickly mix

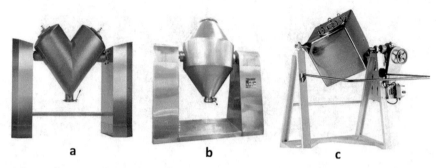

Fig. 2.5 V type (**a**), cone type (**b**), and cube type (**c**) tumbling mixer. (Yenchen Machinery Co., Ltd., Taiwan)

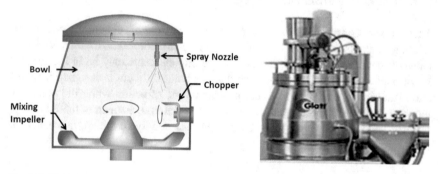

Fig. 2.6 High shear mixer granulator. (Copyrights: Chung Jin Tech, South Korea)

the powder. In case granulation is required simultaneously, granulating agent is added and the impeller speed is lowered, as shown in Fig. 2.6.

2.4.7 Equipments for Mixing of Semisolids

For mixing of semisolids, different mixers like planetary mixers and sigma blade mixers are commonly used.

2.4.7.1 Planetary Mixer/Planetary Bowl Mixer

The planetary mixer has a bowl and a vertically mounted mixing element attached to a shaft that is placed off-center and holds a spinning arm (Fig. 2.7). The paddle thus moves around the mixing bowl while rotating around its own axis. The rotation of mixing element around the bowl resembles the movement of planets revolving around the sun, which is why this mixer is called planetary. It is commonly used as domestic mixer in kitchen. In the pharmaceutical industry, it is used for mixing of

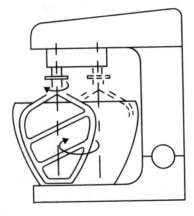

Fig. 2.7 Planetary mixer. (Courtesy: Grydle and Sync Pvt. Ltd., India)

solids and semisolids. It is important to repeatedly scrape off the material from the paddle.

References

1. Cameron AT. Granular Material Mixing: Experiments for Calibration and Validation. The Journal of Purdue Undergraduate Research. 2018;8(1):2–9.
2. Cohen EM, Lin LY. Active Pharmaceutical Ingredients. Generic Drug Product Development: solid oral dosage forms (Drugs and the Pharmaceutical Sciences). 2004; 17–30.
3. Cullen PJ, Romañach RJ, Abatzoglou N, Rielly CD. Pharmaceutical blending and mixing: John Wiley & Sons; 2015 : 479-490.
4. Challener CA. Optimizing the Selection of Mixing Equipment, https://www.pharmtech.com/view/optimizing-selection-mixing-equipment-0
5. Brennan JG. Mixing, emulsification and size reduction. Food processing handbook, Second Edition. Wiley-VCH Verlag GmbH & Co. KGaA, 2011:363–406.
6. Bordbar A, Taassob A, Kamali R. Diffusion and Convection Mixing of Non-Newtonian Liquids in an Optimized Micromixer. The Canadian Journal of Chemical Engineering. 2017;96:1829–1836. https://doi.org/10.1002/cjce.23113
7. Leonard A. Overview of Turbulent and Laminar Diffusion and Mixing. Analysis and Control of Mixing with an Application to Micro and Macro Flow Processes. Springer, Vienna. 2009: 1–33.
8. Asachi M, Nourafkan E, Hassanpour A. A review of current techniques for the evaluation of powder mixing. Advanced Powder Technology. 2018;29(7):1525–49.
9. Hogg R. Mixing and Segregation in Powders: Evaluation, Mechanisms and Processes. Kona Powder and Particle Journal. 2009;27:3–17.
10. Tunuguntla DR, Weinhart T, Thornton AR. Comparing and contrasting size-based particle segregation models. Computational Particle Mechanics. 2017;4(4):387–405.
11. Velázquez C, Florían M, Quiñones L. Chapter 19 - Monitoring and control of a continuous tumble mixer. In: Singh R, Yuan Z, editors. Computer Aided Chemical Engineering. 41: Elsevier; 2018: 471–87.
12. Brone D, Muzzio F. Enhanced Mixing in Double-Cone Blenders. Powder Technology. 2000;110:179–89.

Chapter 3
Comminution

Faiza Hanif and Majeedullah

Abstract Particles with dissimilar size(s) and size distribution demonstrate diverse behaviors. Smaller particles have more surface area and their bulk density, porosity, flow, and solubility considerably vary from large particles. All these properties will influence the formation, packaging, and processing of dosage forms. Thus, comminution is considered a key step in manufacturing of pharmaceutical products. This chapter discusses the concept of size reduction, its importance in pharmaceutical processes, mechanisms involved, and various factors that affect size reduction. It also describes different techniques of particle size analysis. Moreover, the design features and working principle of the equipments used for size reduction of solids, dispersions, and semisolids are also presented in detail.

Keywords Particle size reduction · Size distributions · Cutter mill · Hammer mill · Oscillating granulator · Fitz mill · Triple roller mill and colloid mill

3.1 Introduction

The reduction of bigger particles into smaller ones by the application of external (mechanical) force(s) is called comminution [1].

Pharmaceutical raw material may contain lumps that cannot be processed as such. For these materials, size reduction is an inevitable step. The term comminution/milling/grinding is used in the context of size reduction of solid material, while the relevant term for size reduction of liquid material (droplets) is emulsification and atomization [2].

F. Hanif · Majeedullah (✉)
Department of Pharmacy, Kohat University of Science and Technology, Kohat, Pakistan
e-mail: drmajeed@kust.edu.pk

© The Author(s), under exclusive license to Springer Nature Switzerland AG 2022
S. A. Khan (ed.), *Essentials of Industrial Pharmacy*, AAPS Advances in the
Pharmaceutical Sciences Series 46, https://doi.org/10.1007/978-3-030-84977-1_3

3.2 Importance of Particle Size

Particle size can affect many physicochemical properties of dosage forms. As, in pharmaceutical suspensions, the rate of sedimentation and redispersion of the sediment is greatly affected by particle size. Therefore, particle size needs to be optimized for stable formulation [3]. Similarly, in tablets and capsules, the flow, compaction, and compressibility of powder also depend on particle size [4]. Moreover, particle size can affect texture, aesthetic appearance, and spreadability of semisolids. Particle size is also critical for irritability of ophthalmic semisolid products [5].

3.3 Mechanism of Size Reduction

When an external mechanical force is applied to a material, so that the cohesive forces in a matter are exceeded, the material is fragmented into smaller particles [5]. There are several mechanisms involved in size reduction, such as cutting, compression, impact, and attrition (summarized in Table 3.1) [1].

3.4 Factors Affecting Size Reduction

The following are some of the important factors that affect the size reduction process.

Hardness and Toughness Hardness is the ability of the substance to withstand pressure, and toughness is the ability of a substance to resist being fractured

Table 3.1 Mechanism of size reduction

Mechanism of size of reduction	Diagram	Principle
Cutting		Material is cut by means of sharp blades, e.g., scissors, kitchen blender
Compression		Material is crushed by application of pressure, e.g., nut crackers
Impact		Material is forcefully hit by a moving object, or the material strikes a stationary surface at high speed, e.g., hammer
Attrition		Material is subjected to pressure as in compression, but the surfaces are moving against each other, e.g., levigation by mortar and pestle

when pressure is applied. It is easy to break soft materials to small sizes than hard materials. Similarly, a tough substance is also very difficult to break [6].

Abrasiveness is the ability of a material that causes it to be worn out when friction is applied. If the grinding material is abrasive, the chances of contamination with metal particles will be high due to worn out of metal during processing [7].

Stickiness is the ability of a material to stick to the surfaces of the grinding mills or sieve surfaces. It causes difficulty in size reduction process. Increased stickiness causes the increased adherence of the material to the grinding mills resulting in choking of the mesh [7].

Softening with Temperature Heat generated during the size reduction process can lead to softening of some waxy drugs or stearic acid. This can be avoided by cooling the mill [7].

Structure Some materials have unique structures which can affect size reduction process. For instance, vegetable drugs with cellular structure often change to long fibrous materials when pressure is applied, while minerals that have lines of weakness are reduced in size, producing flake-like particles [8].

Moisture Content The presence of moisture can affect hardness, toughness, and softness of material. The material for size reduction should either be dry or wet but should not be damp [9].

Feed Size should be kept optimum in order to avoid choking of the mill [7].

3.5 Particle Size Distribution

Naturally particles exist in variable sizes and shapes. This size difference varies from smallest particles to the largest ones. There is no specific method to describe particle size of each and every particle in a powder sample [10]. Nevertheless, the size can be described using statistical methods that present particle size in a single dimension by different ways [6]. The tabular form is the most appropriate and precise method. Besides, frequency as a function of size, bar graph, histogram, etc. are also easily understandable representations of particle size distribution. An example is given in Fig. 3.1 [11].

3.6 Particle Size Analysis

Particle size analysis is performed to determine particle size distribution in a given powder.

Fig. 3.1 Histograms of
cumulative size distribution

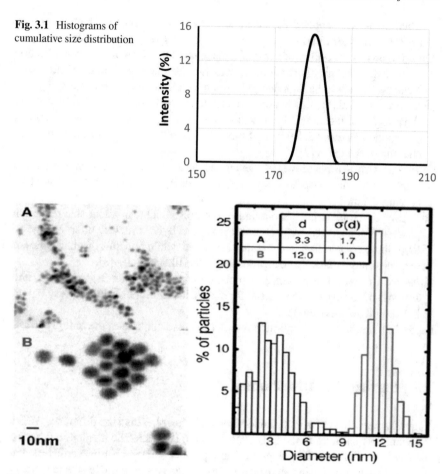

Fig. 3.2 Size analysis by microscopy

3.6.1 *Microscopy*

The direct method for size measurement is the microscopy. A Calibrated filar
micrometer eyepiece is used to measure the diameter of the particles. The results are
presented graphically, as shown in Fig. 3.2. The lower limit of microscopy can be
enhanced by using lenses of different resolutions, for instance, particles in nanometer size range can be measured using scanning electron microscopy (SEM), scanning probe microscopy (SPM), etc. [10].

3.6.2 Sieving

Sieving is the most economical and simplest method for measuring the particle size of a powder. However, the size measured by sieving method is not very precise [10]. The lowest size that can be measured by this method is 50 μm, or 10 μm in case of micromesh sieves. The device comprises rack of sieve pans sequentially arranged such that sieve having smallest pores is placed on top while the sieve with largest pores is placed at the bottom. A certain amount of powder is placed over the upper pan and shaken mechanically. The residual powder left over each sieve pan is weighed [7].

Sieving is affected by factors, such as sieving time, the powder load, and the type of motion during shaking. Therefore, all of these factors need to be optimized.

Mesh number is a numerical number of holes per linear inch on a sieve. There are different series of mesh numbers. Table 3.2 shows the standard sieve numbers and the corresponding openings expressed in microns.

3.6.3 Sedimentation

Sedimentation refers to the settling down of particles in a medium. This method can be used for particles in the size range of 1–200 μm. Moreover, it can only be applied to dilute dispersion where the solid concentration is not more than 2% w/w [10]. The results are expressed as size weight distribution curve. It is based on the sedimentation rate which itself is reliant on the particle sizes, according to Stoke's equation:

$$d = \sqrt{\frac{18\eta h}{(\rho_s - \rho_i)gt}} \qquad (3.1)$$

where d is the diameter of particles, η represents viscosity of the medium, h is the distance of fall in time t, g denotes gravitational constant, ρ_s is the density of particles, and ρ_i represents density of medium.

Table 3.2 Mesh number and opening of standard sieves

Mesh number	Sieve opening in μm	Mesh number	Sieve opening in μm
2.00	9500	70	212
3.5	5600	80	180
4.00	4750	100	150
8.00	2360	120	25
10	2000	200	75
20	850	230	63
30	600	270	53
40	425	325	45
50	300	400	38
60	250		

Source: USP 31-NF 26

Fig. 3.3 Andreasen
apparatus

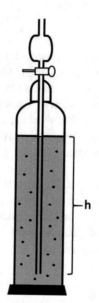
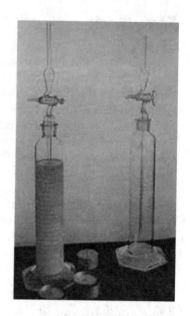

This method is also called *Andreasen method*; briefly, the sample powder is dispersed in a suitable medium (1% w/v). The suspension is placed in Andreasen pipette, shown in Fig. 3.3. At fixed time points, a small portion of sample (10 mL) is withdrawn from the medium and dried without affecting the suspension. The diameter of particles is determined using Stokes equation. Particles of different sizes have different sedimentation rates, i.e., the larger particles will settle down very fast.

3.6.4 Conductivity Method

Conductivity method is also known as Coulter principle and it analyze sizes of individual particles. It requires only trivial amount of sample for analysis and is one of the common method used for particle size measurement. This method relies on fluctuation in electric current caused by particles [12]. An evenly dispersed suspension is formed in an electrolytic solution. A tube having a small orifice of squat path length is immersed in the suspension, and an electrode is placed on both sides of the orifice, as shown in Fig. 3.4. A pump creates movement of electrolyte through the orifice, giving a conductive path in between the two electrodes. A slight electrical current is established in between the electrodes. Individually, electrolyte and particle (s) pass through the orifice. The particle, being non-conductive, hinders the electrical current movement, as they move in the orifice. This generates an electrical signal proportionate to the volume of the particle. Every single particle is measured and categorized according to the volume, thus constructing a volume frequency

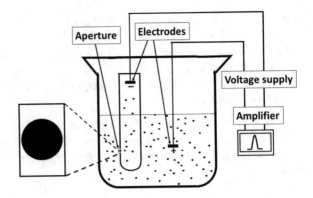

Fig. 3.4 Coulter counter

distribution. The particles are considered to be spherical and a particle diameter can be determined from volume [11].

3.7   Equipments Used for Size Reduction

Selection of size reduction equipment depends on the nature of material to be micronized and required particle size. Each mill gives particles of different sizes and shapes. Once the size range is established (i.e. coarse, intermediate, or fine), then the method can be easily selected [13].

3.7.1   Size Reduction of Powder

Laboratory-scale size reduction done in pestle and mortar is called *trituration*. For industrial-scale size reduction, a variety of equipments are used, called *mills*. These mills typically consist of three basic parts, i.e., feed channel, grinding mean (rotor), and discharge channel.

Feed Channel: The material is introduced into the mill through feed channel.
Grinding Mean: The size of material is reduced by rotating blades or other means of grinding.
Discharge Channel: gives off the reduced particles.

The feed rate and discharge rate should be equal. Gravitational force is typically responsible for material to be discharged from the mill. However, for ultrafine powders, gravitational force alone is not sufficient so air or inert gas is used in order to move particles out of the mill. The milled particles are either coarse, intermediate, or fine [12].

Coarse (No. 20): All particles pass through mesh number 20 and not more than (NMT) 40% pass through mesh number 60.

Intermediate (No. 40): All particles pass through mesh number 40 and NMT 40% pass through mesh number 80.

Fine (No. 60): All particles pass through mesh number 60 and NMT 40% pass through mesh number 100 [14].

The milling process where the material is reduced to the desired size by moving once through the mill is called *open circuit milling*. Conversely, if large particles are first reduced and then transferred to the grinding chamber it is called *closed circuit milling*.

3.7.1.1 Cutter Mill

Design Features Cutter mill has a rotor that has 2–12 knives attached to it. The speed of rotation can vary from 200–900 rpm. It also has stationary knives fixed to the wall of milling chamber. An adjustable screen is fixed at the bottom of milling chamber for controlling particle size, as shown in Fig. 3.5.

The size reduction involves the cutting mechanism. The resultant product is coarse. Particle size can be tuned by adjusting the rotor size, the distance between two adjacent knives, and by changing the screen [12].

Uses: It is frequently used to reduce particle size of crude drug before extraction. It is common for crushing of crude drug materials that are hard and tough in nature. Moreover, it can also be used to mill dried granules [15].

Advantages: It is simple and cheap mill which does not require any complex operation procedure. It can be used for materials that cannot be reduced by other methods.

Limitations: Material cannot be reduced to fine particles, and the produced particles are coarse in size. Cutter mill is difficult to clean after use.

Fig. 3.5 Cutter mill. (Courtesy: 911 Metallurgy Corp., Canada)

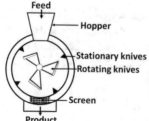

Fig. 3.6 Hammer mill
(Image source: www.
commons.wikimedia.org)

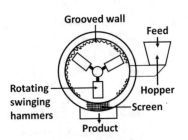

3.7.1.2 Hammer Mill

Design Feature It is also called impact mill, since the mechanism involved is impact. Hammer mill consists of a high-speed rotor to which swinging hammers are attached. It has a feed inlet at the top or center, through which drug material is fed into the chamber [16]. For particle size adjustment, there is a perforated screen above the discharge outlet as shown in Fig. 3.6.

Operation: The material is fed into the main chamber through inlet. The material is hit by swinging hammers that are attached to a shaft that rotates within the chamber at high speed (7600 rpm/min). The substance is crushed by repeated impact action of the hammers and collisions with the grooved grinding chamber walls. The perforated metal screens retain coarse materials for further grinding while small-sized particles pass through. Brittle material is best reduced by blunt hammers, whereas fibrous material by cutting edges [15]. The final size is controlled by hammer speed and the hole size of screen.

Uses: Hammer mill can be used for size reduction of almost all types of materials (barks, leaves, roots, crystals, filter cakes); therefore, it is widely used in pharmaceutical industry. Besides, hammer mill is also used for the size reduction of both wet and dry granulation excipients prior to formulation of compressed tablets [16].

Advantages: It is a rapid action mill and can be used for grinding of different types of materials. Since, no hammer surfaces move against each other, therefore, there are less chances of metal detachment from the equipment and hence less chances of contamination.

Limitations: It is not suitable for heat-sensitive materials because at high operation speed it generates heat. There is a need for optimization of the rate of feed to avoid choking of the mill [12].

3.7.1.3 Ball Mill

Design Feature: Ball mill has a horizontally rotating chamber whose length is rather greater than its diameter and is either filled with stainless steel or porcelain balls. The ball charge is stated as % age of the chamber filled by the balls. Normally balls cover 30–50% of the mill volume. The outlet is covered with coarse screen to prevent loss of the balls [12]. There are different variations of these mills. For instance,

Fig. 3.7 Ball mill.
(Courtesy: Sepor,
Inc., USA)

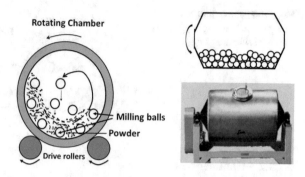

in *pebble mill*, the grinding material is pebble. When the grinding material is rod, then it is called *rod mill*. In *tube mills* the length of balls is four times greater than its diameter [15].

Operation: The chamber rotates horizontally that reduces particle size by both attrition and impact. If the balls rotate at slow speed, it provides attrition effect on the powder, while at high rotation speed the ball provides impact action [12], as shown in Fig. 3.7.

Advantages: Ball mill can provide fine grinding. It can be used for both dry and wet milling. Toxic materials can be grinded in ball mill safely because it is a closed system. It is an economical equipment and easy to install and operate. It is effective for hard and abrasive material [6].

Disadvantages: It produces very high noise.

3.7.1.4 Oscillating Granulator

Design Feature: It consists of a hopper, granulating rods, sieves, and oscillating rotor. All parts are made from stainless steel. It is designed such that the screen can be held and stretched through specific arrangements and a uniform gap is maintained throughout the operation [12], as shown in Fig. 3.8.

Operation: It has one rotor with five rods attached that oscillates at about 180 rpm on its horizontal axis. The material is feed into the hopper which then falls on the oscillating rods. The powder passes through the sieve and is collected [6].

Uses: It is used for homogenization and size reduction of powders and granules in pharmaceutical industries. It is used for dry and wet granulation of pharmaceutical materials.

Advantages: It is made up of stainless steel and chances of contamination are less. Its installation is simple and it is easy to operate [12].

Limitations: Dust generation is common during crushing process.

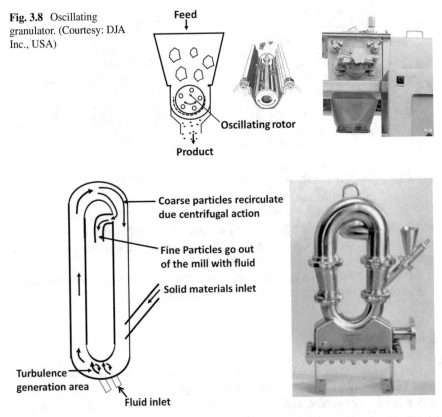

Fig. 3.8 Oscillating granulator. (Courtesy: DJA Inc., USA)

Fig. 3.9 Fluid energy mill. (Courtesy: International Chemical Scientific Co., Ltd., South Korea)

3.7.1.5 Fluid Energy Mill

Design Feature: It consists of an inlet for feed, outlet for discharge of comminuted material, loop of pipe with 20–200 mm diameter and 2 m high, and a series of nozzles for air and an inert gas, as shown in Fig. 3.9.

Operation: It involves both impact and attrition mechanisms for size reduction. The material is suspended in high velocity air that is introduced through the nozzles (100–150 lb/Inch^2 (psi)). The particles impact the walls of the grinding chamber and other particles. The collision with the walls and the interparticle attrition breaks the particles into smaller pieces [12]. These particles are then collected into the cyclone collector, which is attached with the discharge outlet. The large particles are centrifugated and recycled into the grinding chamber for further size reduction. The continuous particle collision can yield particles up to about 5 μm size [6].

Uses: Fluid energy mill is the method of choice for drugs that require high level of purity. Moreover, it is also favorable for heat-sensitive material [12]. Fluid energy mill is commonly used for the size reduction of alumina, kaolin, zinc, etc.

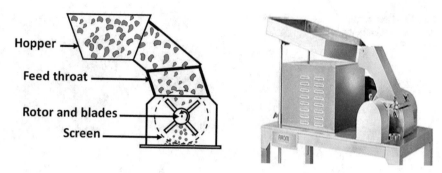

Fig. 3.10 Typical Fitz mill. (Courtesy: Pharmao Industries Co., Ltd., China)

Advantages: Fluid energy mill produces very fine powder. It has no moving parts so wearing of the parts is minimal; hence chances of contamination are less. It is suitable for thermolabile (heat-sensitive) substances [6]. Moreover, the mill is easily cleanable.

Limitations: Particles may tend to aggregate after the size reduction. Fluid energy mill has a high energy impact so it makes amorphous content [15].

3.7.1.6 Fitz Mill

Design Feature: It consists of feed throat. Several designs of feed throat are available. The type, number, and design of the blade actually control the level of reduction attained. The screen and rotor speed both are adjusted according to desired particle size [17].

Operation: The material enters into the grinding chamber and is reduced into smaller particles by high-speed rotating blades, as shown in Fig. 3.10. Thus, it works on the principle of cutting.

Uses: Fitz mills are the standard mills for the pharmaceutical industry because of their accuracy and precision in particle size reduction. It is also used in continuous batch production and research.

Advantages: The end results of Fitz mill are accurate and predictable. The equipment is easy to clean and is simple in operation. It can be scalable to larger units (Table 3.3).

3.7.2   Size Reduction of Pharmaceutical Dispersions

Dispersion is a heterogeneous two-phase system in which internal phase is dispersed in the continuous (external) phase or vehicle. In case of emulsions/creams, two immiscible liquids are blended to disperse one into the other, while in case of suspension/pastes, solid particle (s) are dispersed in a liquid by mixing [18].

Table 3.3 Comparison of different mills in pharmaceutical particle size reduction

Mill	Mechanism of action	Size (mesh no.)	Advantages	Disadvantages
Cutter	Cutting	20–80	Used for fibrous crude animal/vegetable drugs	Cannot be used for friable materials
Revolving	Attrition and impact	20–200	Abrasive materials can be finely grinded	Cannot be used for soft materials
Hammer	Impact	4–325	Used for almost all drugs	Cannot be used for abrasive materials
Roller	Pressure	20–200	Used for soft materials	Cannot be used for abrasive materials
Attrition	Attrition	20–200	Used for soft and fibrous materials	Cannot be used for abrasive materials
Micronizer	Attrition and impact	1–30 μm[a]	Used for moderately hard and friable materials	Cannot be used for soft and sticky materials

[a]Particle size expressed in micrometer

The emulsion preparation involves energy input to create a homogeneously dispersed droplets [1]. In suspension preparation, the first step is to get the right sized particle usually in the micrometer range. Particle size reduction of the dispersed phase is attained by dry milling prior to incorporation into the dispersion medium [10] or milling of the final product to obtain uniform size particles. Size reduction in emulsions and suspensions is accomplished by a number of different types of agitators and mills.

3.7.2.1   Agitators

Conventional shaking or agitation can effectively be used to make emulsion. The equipment is normally used for emulsification of easily dispersible, low-viscosity oils. It is also used for dispersion of solids in liquids. Containers using rotating impellers are employed for agitation (detailed in Sect. 2.3.1).

3.7.2.2   Mechanical Mixers

In propeller and/or impeller mixers, propeller is connected to a shaft run by an electric motor. This type of mixers can perform both functions of stirring and emulsification. They are used for preparing emulsions with low viscosity [19]. Turbine mixer has additional blades (with or without a pitch) that offer more shear than propellers and is suitable for emulsion preparation of both low-viscosity and medium-viscosity consistencies [1] (detailed in Sect. 2.3.1).

Heavy-duty mixers are used for high-viscosity emulsions, where the mixing regime changes from turbulence (as in agitators) to one wherein viscid drag forces dominate. Also, some mixtures display non-Newtonian behavior (fluids whose flow is not defined by a single constant viscosity value), and mixing of such mixture demands superior heavy-duty mixers like double arm, planetary, and dual and triple shaft mixers [17].

3.7.2.3 Colloidal Mill

Colloidal mill is often used for manufacture of colloidal solutions or very fine emulsions and suspensions. It also gives smooth texture to semisolid final product when used as a final step.

Design Feature: It consists of cone-shaped rotor and stator that are 0.005–0.075 cm apart from each other. The rotor is coupled with the motor which rotates at high speed (3000–20,000 rpm). The rotor/stator is cone-shaped and has three stages of increasing grooves that changes direction in each stage for increased turbulence. A hopper and a discharge chamber are also part of colloid mill. The stator is adjusted to get the desired gap with the rotor, as shown in Fig. 3.11.

Operation: The material is introduced into the rotor through hopper which is then thrown onto the stator by centrifugal force. The rotor operates at a very high speed that generates hydraulic shear force that cuts the particles to smaller size. The product is cleared through an outlet and may be recirculated.

Uses: It is used for the size reduction of dry materials. It is also used to reduce particle size of suspension and the globule size of emulsion [12].

Advantages: Colloidal mill can reduce the particle size to 1 μm or less. It can reduce particles in the presence of liquid. The particles reduced are used in preparation of suspension, ointment, cream, and lotion [6].

Limitations: Liquid is used in colloid mill so there is chance of contamination of the product. It cannot be operated continuously because it requires a lot of energy to operate [12].

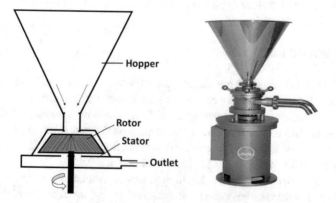

Fig. 3.11 Colloid mill used for size reduction of suspensions and emulsions

3.7.2.4 Homogenizers

Homogenization simply refers to reduction of globules size in emulsion or particle size in suspension, so that each dose has uniform composition [1].

Design Feature and operation: Simple homogenizer has a pump to rise the pressure of the dispersion (500–5000 psi) and an orifice through which the fluid strikes upon the homogenizing "valve," which is held on valve seat by strong spring, as shown in Fig. 3.12. When the pressure rises, some of the dispersion leaks between valve and valve seat, and this subjects the product to extreme turbulence and hydraulic shear [1]. For best emulsification, two-stage homogenizers are built, where emulsions receive second treatment directly after first run [15]. *Silverson mixer* (rotor/stator mixer) is often classified as a homogenizer; it can produce droplet size of 2–5 μm [20]. *High-pressure homogenizer* (HP) is comprised of a displacement pump that pumps liquid into the valve area at great pressure (50–500 Mpa) and an adjustable gap width that is kept 5–20 μm [21]. When a dispersion is passed under tremendous pressure through a narrow orifice, the size of dispersed phase is reduced [20]. *Piston homogenizer* (PH) has a powerful positive displacement piston-type pump to yield high pressure (3000–10,000 psig) to force the coarse mass through a designed restricting wall, where very high shear loads are applied. Here both *turbulence* and *high shear* forces are the key factors for reducing sizes of globules. It has continuous yield capacity of 2500 Lh^{-1} at 15 hp to 50,000 Lh^{-1} at 150 hp [21].

Uses: APV Gaulin homogenizer (a simple homogenizer) is most frequently used in liquid emulsion preparation. The HP homogenizer is used in the manufacture of emulsion preparation, microemulsions, and particle size reduction [20]. HP type is now considered the most powerful homogenizer for making emulsions and suspensions [11, 21].

Advantages: A homogenizer does not permit air entry into the final product.

Limitations: PH cannot handle the feed mixture over 200 cps. It has high maintenance cost and the product lacks consistency and has batch-to-batch variability [20].

Fig. 3.12 Homogenizer

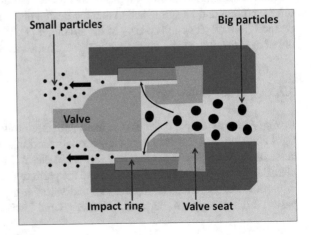

Small particles

Big particles

Valve

Impact ring

Valve seat

Inline Mixers: Many mixers, such as Silverson mixer, colloidal mill, piston homogenizer, and ultrasonic vibrating homogenizer are now designed to offer a range of in-tank and inline mixing, and are extensively used for preparation of emulsions, from pilot scale to bulk production units.

3.7.2.5   Ultrasonic Devices

For low-viscosity emulsions use of ultrasonic vibration has been reported that causes compression and refraction in different regions of liquid that result in high shear. These devices are not practical for bulk-scale production of emulsions [1].

3.7.3   Size Reduction in Semisolids

Semisolids as a dosage form include creams, gels, pastes, and ointments. Their industrial processing is very similar; hence this topic will cover all semisolids collectively [1]. Equipment for semisolids is built to perform certain unit tasks, such as milling, separation, mixing, emulsification, and deaeration. *Separators* are used to separate materials of different sizes, shapes, and densities, e.g., centrifugal separators or vibratory shakers [11]. Mixing of the active constituents and other formulation ingredients with the ointment base is achieved by different types of agitator mixers and high-shear mixers. Mixers with heating supplies are also used to help in the melting of bases and mixing of constituents [12].

3.7.3.1   Agitators

In manufacture of semisolids, different types of agitator mixers are used, for instance, sigma blade mixers and planetary mixers. The agitator arms are so built to offer a pulling and kneading action and the design and drive is such that content is cleaned from all sides and corners of the container [1]. Planetary mixer is an example of agitator mixer used for semisolids (see Chapter 2, Section 2.4.7.1).

3.7.3.2   Triple Roller Mill

Design Features: Triple roller mill is fixed with three rollers which are built from a rigid abrasion-resistant material. They are mounted in such a way that they lie in close interaction with each other and revolve at dissimilar speeds [6]. The material placed in between the rollers is crushed and the particle size is reduced. The decrease in particle size is influenced by the gap and variation in roller speeds. Material is passed through hopper A, between the rollers B and C where it is comminuted. Then the material is passed between the rollers C and D where it is further homogenized to obtain a smooth mixture.

Advantages: Triple roller mill offers possibility of continuous operation. Mixing efficiency achieved is very high in terms of content uniformity. Moreover, it provides excellent control of the product temperature [15].

Limitations: Washing of the mill after use is difficult. Overall cost is very high. Chances of contamination of the final product are maximum [6].

3.7.3.3 Vacuum Emulsifying Mixers

It is the most common and extensively used emulsifying mixer nowadays.

Design Feature: Vacuum emulsifying mixer has three distinct mixers, a central emulsifying pot, pre-treating oil pot, water pot, and work frame [22].

Operation: The oil-phase constituents in oil pot and the water-phase constituents in water pot are vigorously mixed before being dropped in the central emulsifying mixer. This central pot also has a rotor/stator that pulls, mixes, shears, and forces the constituents together. After thorough mixing in central pot, the temperature is raised to promote further solubility. This mixture is then removed with hydraulic pumps and cooled [22].

Uses: It is the best emulsifier for emulsification of viscous materials with high solid content like cream, ointment, lotions, and gels [22].

References

1. Wang YB, Williams RO. Powders. Remington Essentials of Pharmaceutics. London : Pharmaceutical Press. 2013: 411–432.
2. Loh ZH, Samanta AK, Sia Heng PW. Overview of milling techniques for improving the solubility of poorly water-soluble drugs. Asian Journal of Pharmaceutical Sciences. 2015;10(4):255–74.
3. Papuga K, Kaszubkiewicz J, Kawałko D. Do we have to use suspensions with low concentrations in determination of particle size distribution by sedimentation methods? Powder Technology. 2021;389:507–521.
4. Chang S-Y, Sun CC. Effect of particle size on interfacial bonding strength of bilayer tablets. Journal of Powder Technology. 2019;356:97–101.
5. Kulkarni V, Shaw C. Particle size analysis: an overview of commonly applied methods for drug materials and products. Essential Chemistry for Formulators of Semisolid Liquid Dosages. 2016: 137–44.
6. Kumar R, Thakur AK, Chaudhari P, Banerjee N. Particle Size Reduction Techniques of Pharmaceutical Compounds for the Enhancement of Their Dissolution Rate and Bioavailability. Journal of Pharmaceutical Innovation. 2021:1–20.
7. Sushant S, Archana K. Methods of size reduction and factors affecting size reduction in pharmaceutics. International Research Journal of Pharmacy. 2013;4(8):57–64.
8. Opatová K, Zetková I, Kučerová L. Relationship between the Size and Inner Structure of AM Powder Particles. 2018. doi:10.20944/preprints201811.0453.v1.

9. Ahlneck C, Zografi G. The molecular basis of moisture effects on the physical and chemical stability of drugs in the solid state. International Journal of Pharmaceutics. 1990;62(2–3):87–95. DOI:10.1016/0378-5173(90)90221-O

10. Buanz A. Powder characterization. In: Adejare A., (Ed), Remington the Science and Practice of Pharmacy, Elsevier, Amsterdam, Netherlands. 2021: 295–305.

11. Yadav KS, Kale K. High pressure homogenizer in pharmaceuticals: understanding its critical processing parameters and applications. Journal of Pharmaceutical Innovation. 2019:1–12.

12. Aulton ME, Staniforth JN. Particle size reduction and size separation. Aulton's Pharmaceutics E-BOOK: The design and manufacture of medicines, 4th edition, 2013. 156–169.

13. Duroudier J-P. Grinding: Principles and Theories. Size Reduction of Divided Solids, 1st edition, Elsevier; 2016.

14. Brown W, Marques MR. The United States Pharmacopeia/National Formulary 2013: 319.

15. Seibert KD, Collins PC, Luciani CV, Fisher ES. Milling operations in the pharmaceutical industry. Chemical Engineering in the Pharmaceutical Industry: Active Pharmaceutical Ingredients. 2019: 861–879.

16. Su D, Yu M, Study of Corn Stover Particle Size Distribution Characteristics for Knife Mill and Hammer Mill. IOP Conference Series: Earth and Environmental Science. 2019; 358(5):052060.

17. Cullen PJ, Romañach RJ, Abatzoglou N, Rielly CD. Pharmaceutical Blending and Mixing. John Wiley & Sons, Inc., Hoboken, New Jersey. 2015: 25–78.

18. Anger CB, Rupp D, Lo P, Takruri H. Preservation of dispersed systems. Pharmaceutical Dosage forms, CRC Press, Boca Raton, Florida. 2020: 377–436.

19. Torotwa I, Changying J. A Study of the Mixing Performance of Different Impeller Designs in Stirred Vessels Using Computational Fluid Dynamics. Designs 2018:2;10. https://doi.org/10.3390/designs2010010

20. Singh SK, Naini V. Homogenization and homogenizers. In: Encyclopedia of Pharmaceutical Science and Technology, Fourth Edition, CRC press, Boca Raton, Florida. 2013: 1848–1854.

21. Yadav KS, Kale KJ JoPI. High pressure homogenizer in pharmaceuticals: understanding its critical processing parameters and applications. Journal of Pharmaceutical Innovation, 2020;15(4):690–701.

22. Garcia EE, Kimura C, Martins AC, Rocha GO, Nozaki J JBAoB, Technology. 2017 High Quality Perforated Panel for Metal Mesh Curtain Wall. Brazilian Archives of Biology and Technology. 1999;42(3):281–90.

Chapter 4
Filtration

Farmanullah and Sajid Hussain

Abstract The prepared pharmaceutical solutions require passage through a thin membrane having many tiny openings. The aim is to eliminate viable particles to clarify the solution. This chapter covers solid-liquid filtration procedures and machinery. This chapter serves as an introduction to the vital concepts and terms a pharmacy student needs to know. An overview of filtration, factors affecting filtration, basic principles and mechanisms, and filter media are all concisely explored. Finally, filtration equipment types have been discussed in detail along with their assemblies and operational mechanisms.

Keywords Filtration · Filter aid · Filtration mechanism · Filtration devices · Filter media

4.1 Introduction

Filtration is the process used to separate solid from fluids (liquids or gases) with the help of a filter medium. Some of the common nomenclature in filtration include:

When a mixture of solid and fluid, called *feed*, is passed through a *filter medium*, i.e., a porous material that provides solid-fluid separation, the solid is retained, called *residue*, while the fluid passes through, called *filtrate* or *effluent*. The process is termed as *cake filtration* when the residue is to be recovered, while it is called *clarification* when the intended purpose is to get clear filtrate.

Farmanullah · S. Hussain (✉)
Department of Pharmacy, Kohat University of Science and Technology, Kohat, Pakistan
e-mail: hussainsajid@kust.edu.pk

© The Author(s), under exclusive license to Springer Nature Switzerland AG 2022
S. A. Khan (ed.), *Essentials of Industrial Pharmacy*, AAPS Advances in the Pharmaceutical Sciences Series 46, https://doi.org/10.1007/978-3-030-84977-1_4

4.2 Factors Affecting Rate of Filtration

In initial phase of filtration process, estimation of the resistance to flow is fairly easy. But if it continues, solid gets trapped in the filter pores that make accurate estimation of filter resistance very difficult. Hence a number of mathematical models are used and are differently applicable to various filtration methods but all follow the same basic rule, i.e., the rate of filtration is directly proportional to the driving force and inversely proportional to the resistance [1], as shown in Table 4.1. Filtration area and pressure difference are the two contributing components of driving force. Thus, filtration rate is directly affected by the area available for filtration and the pressure difference across the filter medium.

Resistance means all the parameters that hinder the filtration process. Resistance is not constant; it increases as particles get deposited on the filter medium. Moreover, resistance is also dependent on the viscosity of fluids; viscous liquids are hard to be filtered [2].

4.3 Mechanism of Filtration

The mechanisms by which particles are retained on the filter medium are diagrammatically given in Fig. 4.1. These include:

Straining as observed in sieving. The pores are smaller than the particles to be separated [4]. Thus, the particles are retained on the surface of the medium.

Impingement when a fluid containing particles passes through the filter. The fluid stream curves around the fibers, and the inertia of particles forces them to deposit on the filter fibers [3].

Table 4.1 Different models for determining the rate of filtration [3]

General rate equation	Darcy's model	Poiseuille's model
$\text{Filtration rate} = \dfrac{\text{Driving force}}{\text{Resistance}}$	$\dfrac{dv}{dt} = \dfrac{kA\Delta P}{\mu l}$	$dv/dt = \dfrac{A\Delta P}{\mu(\alpha W/A + R)}$
	where: dv/dt = rate of filtration A = area of filter medium ΔP = pressure gradient at both the sides of filter medium k = permeability constant for the filter medium μ = viscosity of filtrate l = thickness of the filter medium and cake	where: dv/dt = rate of filtration A = area of filter medium ΔP = pressure gradient at both the sides of filter medium μ = viscosity of filtrate α = average-specific cake resistance W = weight of dry cake R = resistance of filter medium and cake

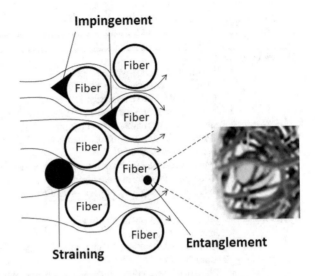

Fig. 4.1 Mechanism of filtration

Entanglement Usually, exhibited when felt is used as filter medium for filtration. The particles are smaller than the pores, and they penetrate into the interior of the filter medium and get entangled in the fibers' mass [5].

4.4 Filter Media

It is the porous material that provides solid-fluid separation. Depending on the retention of solid particles, the filter media may be *surface-type filter media*, i.e., the particles are retained on the surface of filter media, or *depth-type filter media*, where particles move inside the filter media and are retained inside the filter media [6]. The choice of filter medium depends on the desired outcome. It should retain solids without plugging and without excessive bleeding. The ability of a filter medium to remove solid mass from a liquid is called *retention*. The different types of filter media used are summarized in Table 4.2.

4.5 Filter Aid

Filter aids are porous material that are used to improve the porosity of the residue deposited (*filter cake*) on the filter media, thus improving the rate of filtration. Filter aids can be applied by two methods. It can be either used as pre-coat on the filter media to prevent clogging of the pores by fine particles, or it can be mixed with the slurry to increase the porosity of the cake as it forms [7]. Diatomaceous Earth (the skeleton of ancient diatoms), perlite (an amorphous volcanic glass), cellulose, etc. are some of the common filter aids used in pharmaceutical filtration processes.

Table 4.2 Types of filter media [3]

Type	Examples

Woven type	*Filter cloth*
	    Function as surface-type medium
	    Made of natural, synthetic fibers or metals
	    Cotton is the most commonly used material
	    Nylon is superior for pharmaceutical use (no microbial growth, smooth surface, and negligible absorption properties)
	    Teflon is superior (chemically inert, strength and high temperature)
	Wire mesh (stainless steel)
	    Durable, resistant to plugging, and easily cleanable
	    Good surface filter for cake filtration
	    Installed in filling lines of packing equipment

Nonwoven type	*Felt*
	    Fibrous mass mechanically interlocked
	    Function as depth filter
	    Recommended for gelatinous solutions or slurry of fine particles
	Bonded fabrics
	    Binding of textile fibers with resins, solvents, and plasticizers
	    Not widely used in pharmaceuticals
	Kraft paper
	    Usually used in plate and flame filters
	    Offers controlled porosity, limited absorption characteristics, and low cost
	    Support of cloth or wire mesh is necessary
	Membrane filter
	    Thin membrane (usually, 150 µm thick, 400–500 million pores per square centimeter)
	    Prefilteration is needed to avoid clogging
	    Made of various esters of cellulose or from nylon, Teflon, PVC, polyamide, polysulfone, or silver
	    Pore size depends on particles to be removed (0.2 µm is used for sterile filteration)

4.6 Filtration Devices

4.6.1 Filter Leaf Assembly

Design Features Filter leaf assembly is composed of a chamber containing one or several "filter leaves" sequentially arranged. Each filter leaf is composed of a grooved plate or mesh surrounded by a frame of any shape. The grooved plate and frame are surrounded by a filter cloth. The leaves are connected to a common outlet for filtrate.

Operational Mechanism The feed is supplied to the chamber. The filter leaves are dipped in the feed. Pressure is applied to the chamber, as shown in Fig. 4.2. Due to pressure difference (typically ~8 bars), the liquid passes through the filter cloth, enters the troughs of the plate, and eventually passes through the frame (channel) to the outlet [8]. The solid residue is retained on the filter cloth. The filter leaves are washed by immersion in water or by the application of reverse airflow.

Fig. 4.2 Leaf filter and its
internal structure.
(Courtesy: MBL Group,
Malaysia)

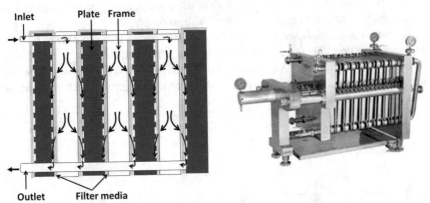

Fig. 4.3 Mechanism of filtration through filter press (www.parksanfilters.com)

4.6.2   Filter Press

Design Features The typical assembly consists of grooved plates, hollow frames, and filter medium (i.e., usually a cloth or filter paper) sandwiched between plates and frames, as shown in Fig. 4.3. A common inlet connects all the frames and the plates are connected to a common outlet [9]. The pressure gauge on the plates is used to monitor the pressure of filtration. The plates and frames are made up of metals, plastics, and sometimes from wood. The wooden plates and frames are kept wet to prevent shrinkage and de-shaping.

Operational Mechanism The plates and frames are set alternatively in a sequence; filter medium is placed in between them. The feed enters the frames via an inlet channel and moves through the filter medium into the plates, where it drains down through the groves to the outlet channel of the plates, which then opens into a common outlet. The filter cake is deposited in frames. Filtration is continued until the frames are entirely occupied by filter cake [10]. As the filtration process extends, the rate of filtration decreases, since the resistance of the cake increases. When filtration rate is substantially lowered, it is preferred to stop filtration operation and remove the cake, rather than continuing the process at very low filtration rate.

4.6.3 Disc Filter Assembly

Design Features It comprises a number of grooved plates and hollow frames. Filter media (circular disc shaped) is sandwiched between plates and frames that are vertically fixed on a central column which has a channel at the circumference. Felt is frequently used as filter media and is usually disposable. The entire housing is closed into a pressure casing. The frames have small holes at the sides that function as inlet for fluid, as shown in Fig. 4.4. Disc filter is compact, portable, and easily cleanable. However, it is mainly used for clarification purpose.

Operational Mechanism When the feed is introduced into the assembly with pressure, it enters the frames through peripheral holes and subsequently passes through filter media. The filtrate moves through the grooved plates and is directed toward the channel at the circumference of the vertical column. The filtrate drains down the channel to the outlet.

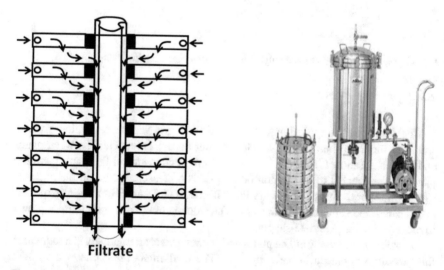

Filtrate

Fig. 4.4 Disc filter assembly internal view (www.parksanfilters.com)

4.6.4 Rotary Drum Filter Assembly

Design Features It consists of several rectangular units having a perforated curved surface, all joined together to form a drum. When rotated, the segments pass through various zones in a continuous succession [11]. A filter cloth covering the surface serves as filter medium. All segments are joined to a rotating duct at the center of the drum (shown in Fig. 4.5). It is mainly used for cake filtration, hence also called rotary drum dryer.

Operational Mechanism The lower surface of the drum is dipped into the slurry. This zone is called pick-up zone or filtration zone. As the drum is rotated the solid is deposited on the surface while the filtrate drains down the segments into the central rotating outlet [12]. In washing zone, a shower of water washes the cake at the surface of the drum, and after washing it is subjected to drying (drying zone). The dried cake is usually scraped with sharp knives (removal zone).

Some filtration operation might need special considerations, for instance, if the cake shrinks upon drying, it is slightly compacted with the help of compression rollers in order to ensure efficient washing. Similarly, scrapers for removal of cake might occasionally pose problems; a string discharge is added. The cake is formed over the strings that is detached when the strings pass through small rollers to get a sharp bend, thus releasing the cake from the surface.

Advantages It has large surface area and offers continuous filtration. The speed can be varied according to the thickness of the cake.

Limitations The equipment is complex and expensive. It is less effective for solids of impermeable cake.

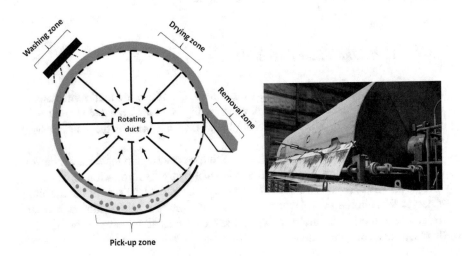

Fig. 4.5 Rotary drum filter (www.nfm-filter.com)

Fig. 4.6 Bag filter
assembly (www.
waterdropfilters.co.za)

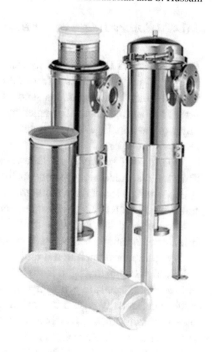

4.6.5   Bag Filter Assembly

Design Features It consists of a nylon bag held in a stainless steel or high-quality plastic cage that is enclosed in a stainless-steel housing.

Operational Mechanism The feed is introduced from the top into the filter housing, it passes through the filter bag, the cake (residue) is left in the bags, and the liquid (filtrate) drains down to the outlet, as shown in Fig. 4.6.

4.6.6   Cartridge Filter Assembly

Design Features It utilizes cartridges for filtration. Cartridges are made with disposable or cleanable filter media that is either string wound or pleated wound on a plastic hardware [13]. The receptacles (cavities) in the base of the housing hold many cartridges, as shown in Fig. 4.7.

The cartridge size, the housing design, and the construction material depend on the intended application. Cartridge filters are mainly used for filtration of water.

Operational Mechanism Feed enters through the inlet at the side of the housing; it passes through the thick layer of filter medium into the central tube of the cartridge and subsequently drains down to the receptacles at the base of the housing that leads to a common outlet.

References

1. Nagy B, Szilágyi B, Domokos A, Tacsi K, Pataki H, Marosi G, et al. Modeling of pharmaceutical filtration and continuous integrated crystallization-filtration processes. Chemical Engineering Journal. 2021;413:127566.
2. Kahshan M, Lu D, Abu-Hamdeh NH, Golmohammadzadeh A, Farooq AA, Rahimi-Gorji M. Darcy-Brinkman flow of a viscous fluid through a porous duct: Application in blood filtration process. Journal of the Taiwan Institute of Chemical Engineers. 2020;117:223–30.
3. Khar RK. Lachman/liebermans: the theory and practice of industrial pharmacy: Cbs Publishers & Distribu; 2013.
4. Tien C. Principles of filtration: Elsevier; 2012.
5. Aulton M, Taylor K. Aulton's pharmaceutics: the design and manufacture of medicines. 2013. Elsevier Health Sciences.
6. Sparks T, Chase G. Section 7 - Filter Selection, Process Design, Testing, Optimization and Troubleshooting Guidelines. In: Sparks T, Chase G, editors. Filters and Filtration Handbook (Sixth Edition). Oxford: Butterworth-Heinemann; 2016. p. 383–413.
7. Harcum S. 2 – Purification of protein solutions. In: Abbott A, Ellison M, editors. Biologically Inspired Textiles: Woodhead Publishing; 2008. p. 26–43.
8. Prager G. Practical Pharmaceutical Engineering: Wiley Online Library; 2019.
9. Guerrini L, Masella P, Migliorini M, Cherubini C, Parenti A. Addition of a steel pre-filter to improve plate filter-press performance in olive oil filtration. Journal of Food Engineering. 2015;157:84–7.
10. Civan F. Chapter 12 - Cake Filtration: Mechanism, Parameters and Modeling. In: Civan F, editor. Reservoir Formation Damage (Third Edition). Boston: Gulf Professional Publishing; 2016. p. 295–341.
11. Liu X, Hu Z, Wu W, Zhan J, Herz F, Specht E. DEM study on the surface mixing and whole mixing of granular materials in rotary drums. Powder Technology. 2017;315:438–44.
12. Rotary drum vacuum filter for small-scale industrial processes or pilot trials. Filtration & Separation. 2004;41(8):14.
13. Cartridge filters target applications where the cost of filtration is critical. Membrane Technology. 2021;2021(5):4.

Chapter 5
Pharmaceutical Packaging

Fawad Ali and Inzemam Khan

Abstract Packaging is one of the major sections of the pharmaceutical industry. The role of packaging evolves from time to time due to the new regulatory guidelines and scientific developments. Each product possesses unique properties. Depending upon the product characteristics, the packaging material may differ from product to product. The choice of the packaging material also depends upon how the drug is produced, transported, and stored. The basic purpose of packaging is to provide efficacy, safety, uniformity, and purity and minimize impurities to comply with the standards. This chapter presents sound knowledge about pharmaceutical packaging, most importantly the different types of material used in packaging of pharmaceuticals.

Keywords Primary packaging · Secondary packaging · Tertiary packaging · Blister packaging · Strip packaging · Collapsible tubes

5.1 Introduction

Packaging is one of the most important pharmaceutical parameters and may be defined as the art, science, and technology of wrapping or protecting products for distribution, storage, sale, and use. Wrapping material around the product serves to contain, protect, promote, identify, and make the product marketable and keep it clean [1].

In other words, packaging can be defined as an organized system whose function is to prepare pharmaceutical goods for transport, sale, logistics, warehousing, and

F. Ali (✉)
Department of Pharmacy, Kohat University of Science and Technology, Kohat, Pakistan
e-mail: fawadali@kust.edu.pk

I. Khan
Department of Pharmacy, University of Peshawar, Peshawar, Pakistan

end use safely and effectively. The main function of packaging is to contain and protect the enclosed product for safe transporting to sell it [2].

5.2 Properties of Pharmaceutical Packaging

The suitable packaging, labeling, and storage of all dosage forms are essential for product stability and safe use. The following are some main properties of packaging:

- Packed content should be enclosed within the packaging material.
- Flavor of the packed content should be intact.
- It should be able to show resilience to a large range of temperature.
- It should have sufficient compulsive strength to withstand heat during steralization and to sustain shock and external stress during shipment.
- Good mechanical strength is also a requirement for packaging to bear the pressure of handling, filling, closing, and transportation.
- The shape of the packaging should be attractive and easy to use.
- If the content of the packaging material contains alkali, it should not leach to the content.
- It should be nontoxic, inert, colorless and tastless.
- The container of the packaging material should not favor the growth of microbes.
- Closure should also be nontoxic and chemically stable with container contents as it is part of the container.

5.3 Types of Packages

5.3.1 Primary Package

Primary packaging is the smallest unit of distribution/or use and is defined as the packaging which is in direct contact with the formulation [3]. For example, bottles for syrup, jar for cream, and pouch for powder, syringes, ampoule, flexible bag, etc. The basic aim of primary packaging is to hold, protect, and/or preserve the final product, especially against contamination.

5.3.2 Secondary Package

It holds the primary package and offers additional protection to the product during handling. It also provides detailed information about the product [4]. For example, carton/paper provides additional safety against mechanical and other environmental hazards. Secondary packaging is the outermost covering. The most common

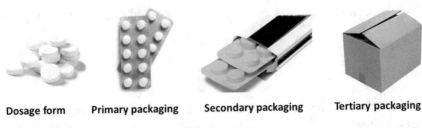

Dosage form   Primary packaging   Secondary packaging   Tertiary packaging

Fig. 5.1 Different types Packagingtypes of packaging

secondary covering is carton. The main composition of the carton is cardboard. The secondary package is not in direct contact with the drug. The secondary package may contain associated components such as a dosing dropper and calibrated spoon.

5.3.3   Tertiary Package

Tertiary packaging is one of the three types of covering which is used to protect, the prepared pharmaceuticals during its shipping and storage. In order to easily transport the bulk of secondary packages, they are often combined and grouped and packed in tertiary packages. The main purpose of the tertiary package is the convenient transport of bulk secondary packages. For example, efficient product shipping is enabled by stretch-wrapped pallets containing several cardboard boxes (Fig. 5.1).

5.4   Common Material for Pharmaceutical Packaging

The selection of packaging material depends on the chemical and physical properties of pharmaceutical product, degree of protection required, compatibility with the dosage form, size and weight of dosage form, filling method, sterilization method to be employed and the cost of product. The most common materials used for packaging of pharmaceuticals are Metals, Polymers, Glass, Plastics, Paper and Rubber. These materials will be discussed in more detail in coming section.

5.4.1   Metal

Metal containers are often used for packaging of non-parenteral pharmaceuticals. Metal is durable, dense, and impervious to moisture and gases. Metal is particularly useful for packaging pressurized containers. Metal containers may be in the form of tubes, blisters, cans and jars, etc. Common metals used in pharmaceutical packaging are:

5.4.1.1 Aluminum

Aluminum is used in pharmaceutical packaging generally as layered material. It is highly customizable. It protects the product from light, oxygen, and moisture. As a result, it ensures longer shelf life of the product. Thickest aluminum is used in preparation of tough containers like aerosol cans, and tubes for effervescent tablets. Aluminum of intermediate thickness are used in preparing collapsible tubes for semi-solids or roll on screw caps while thinnest one is used in flexible foil that are a component of laminated packaging material.

5.4.1.2 Tin

Tin containers are preferred for food, pharmaceuticals and any product for which purity is critical. It is the most chemically inert among the metals used in pharmaceutical packaging. Nowadays, some eye preparations like eye ointments are packaged in purely tin-made containers. The main disadvantage of tin is that it cannot be used for extremely acidic products.

5.4.1.3 Other Metals

Iron is less common in pharmaceutical packaging; nevertheless, tin-coated steel screw caps for jars and aerosols are used. Lead is the most economical metal. However, due to possible toxicity, it is not common in pharmaceutical packaging.

5.4.2 Glass

Glass containers are usually the most preferred material for the packaging of pharmaceutical products. Glass containers are usually employed as primary packages.

5.4.2.1 Composition of Glass

Glass is a three-dimensional network that produces a network by linking silicon atoms with four oxygen atoms in a tetrahedral fashion. Silica (60–80%) and oxides, such as calcium oxide (5–12%), sodium oxide (12–17%), aluminum oxide (0.5–3%), barium oxide, boric oxide, potassium oxide, and magnesium oxide, are the main constituents of glass composition [5]. The presence of silica imparts a high melting point to glass. The oxide composition of glass is the main factor for determining the melting point and viscosity of the glass [6].

5.4.2.2 Classification of Glass

Glass is mainly classified into four main types. The classification of glass is mainly based on the degree of chemical or hydrolytic resistance. The four main types of glasses are type I, type II, type III, and type IV, as explained in Table 5.1 [3].

5.4.2.3 Quality Control Test for Glasses

In order to identify the quality of glass, different quality tests are employed which give us a view of the quality of glass, on the basis of which we select the type of glass for our packaging of our products. These tests are discusses below;

Powered Glass Test is performed at elevated temperature to estimate the amount of alkali leached from a container at a higher temperature. When the glass is powdered, leaching of alkali is increased, which is titrated with 0.02 N sulfuric acid. The indicator used here is methyl red.

Water Attack Test is also performed to determine the leaching properties of the container; for this, it is just treated with soda-lime under controlled humidity.

Table 5.1 Different types of glass used for pharmaceutical packaging

Glass type	Composition	Use
Type I glass (borosilicate glass/neutral glass)	80% silica, 10% boric oxide, a small amount of sodium oxide, and aluminum oxide. Chemically inert and has high hydrolytic resistance due to the presence of boric oxide	Suitable for packaging of parenteral products and non-parenteral products. Can be used for strong acids and alkalis
Type II glass (soda-lime-silica glass/treated soda-lime glass/de-alkalized soda-lime glass)	Obtained by treating type III glass with sulfur. Thus, the hydrolytic resistance is increased. It has a lower melting point compared to type I glass and easier to mold	Suitable for most acidic and neutral aqueous preparations. It can be sterilized before and after filling
Type III glass containers (regular soda-lime glass)	This is an untreated soda-lime glass with average chemical resistance	For powder for parenteral and non-aqueous preparation and non-parenteral preparation packaging.
Type IV glass (general-purpose soda-lime glass)	This type of glass container has low hydrolytic resistance. Not suitable for autoclaving	For topical and oral product packaging

Arsenic Test   is mostly performed for those glasses which are used for packaging of aqueous preparations.

Thermal Shock   is done to observe any change (cracks or breaks) in the structure of the container before or after treated with hot and cold water. Small bottles withstand a temperature differential of 60–80°C [1].

5.4.3   Plastic

"Plastics are group of substances of natural or synthetic origin, consisting chiefly of the polymers of high molecular weight that can be molded into a shape by means of heat and pressure". The entry of plastic as a packaging material occurred in 1950, when different polymers were introduced for commercial use. Nitrocellulose was also discovered in the nineteenth century and was exploited far more quickly. It became one of the first commercially accepted plastics. Today, most pharmaceutical products are packaged in plastic. Plastic bags for intravenous fluids, plastic ointment tubes, plastic film–protected suppositories, and plastic tablet and capsule vials are examples of plastics used in pharmaceutical packaging.

5.4.3.1   Types of Plastics

Basically, two types of plastics, thermoplastics and thermosets, are used for pharmaceutical packaging. The main difference these two is that thermoplastics can be reheated and remolded while thermosets cannot.

Thermosets
These are produced by the polymerization process in which the material is subjected to certain heat and pressure. To obtain better properties, usually fillers and other agents are added to it. Resins, phenolic, melamine, urea, alkyds, epoxide, polyester, polyurethane cross-linked polymer are some of the examples of thermosets. The packaging application include in *Closures, protective lacquer and enamel* (as applied to the outer and inner layer of a metal container) and a range of adhesive systems.

Thermoplastics
Such materials can be softened with heat as they are heat softening materials. It can be heated many times and reshaped to a solid state by cooling. The cross-linked polymerization of branched and unbranched polymers take place in such a process. The molecular structure of plastic plays an important role in defining the properties of plastic.

5.4.3.2   Common Polymers Used for Plastic Containers

Polyethylene

If a barrier against moisture is required, then polyethylene is a good choice, as it provides good resistance to moisture. But its performances to act as a barrier against oxygen and other gases are poor. The four basic characteristics of polyethylene as a packaging material are:

- Stiffness
- Moisture vapor transmission
- Stress
- Cracking

Polypropylene

The main advantage of polypropylene over polyethylene is no stress-cracking at any condition. The polypropylene package may soften if hot aromatic and halogenated solvents or products are placed in it. The good thing with polypropylene is that due to its high melting point such products can be packed in it which are boilable and need sterilization.

Polyvinyl Chloride

Good barrier against gaseous material and stiffness is its main advantage. Polyvinyl chloride quality may be enhanced further if the residual vinyl chloride monomers are reduced in it. To provide shatter-resistant coating, polyvinyl chloride is coated on glass bottles.

Polystyrene

It is rigid and clear crystal plastic. The use of polystyrene for liquid products is not recommended. It has high water and gaseous permeability [7].

EVA (Ethylene-Vinyl Acetate)

Ethylene-vinyl acetate is a copolymer composed of ethylene and vinyl acetate. It has several properties, and these properties depend on these two monomers, i.e., increasing or decreasing the number of these monomers directly affects the properties of ethylene-vinyl acetate. EVAs are soft flexible compounds. Good strength and high elongation even at low temperatures are its main advantages.

EVAC (Ethylene-Vinyl Acetate Copolymers)

EVA is a hard and pliable material. It can be a low-temperature melting rubbery texture-type material or polyethylene-like material. It is mainly used in a hot melt adhesive, for disposable gloves, tubing, and sheeting [1].

5.4.3.3   Additives Used in Plastic Containers

Due to toxicity concerns, the number of additives used in any single plastic are few, and the permissible range for the additive inclusion in plastic to be used for packaging is fairly tough. The main reason is the leaching or extraction of toxic materials from the plastic to the content the container it is enclosing. Plastics may also incorporate processing aids (material) and additives which alter the plastic chemically or physically in some way.

Plasticizers The main purpose of plasticizer is to improve the flow properties of a material and increases softness and plasticity. The content of the plasticizer can be adjusted according to the requirements of the material. As a generalization plasticized PVC usually constitutes at least 20% of plastics. Example of plasticizer include phthalate ester.

Fillers These are inert solid materials. The basic purpose of adding fillers is to reduce the plastic degradation and decrease the cost. Fillers include chalk or calcium carbonate and may be defined as a nonreactive solid substance. Other examples of filler are carbon black (up to 5%), talc, silica, and magnesium carbonate. There are many advantages of the filler, among which one is that it may reduce degradation of the plastic (carbon black may significantly reduce sensitivity to light). The maximum limit of fillers is around a 1:1 ratio. Titanium dioxide, an example of filler, is used up to a 3% level to achieve the desired property.

Stabilizers increase the stability of plastic either during processing or during the molded life of the material. Heat stabilizers are essential to PVC. They have been categorized as used in "polyvinyl chloride" above. Stabilizers are often used to combat the combined effects of heat and light. Example include calcium-zinc salts added to PVC.

Antioxidants Prevent or slow down the oxidation of plastics. Examples of antioxidants include cresols.

Antistatics The main purpose of adding antistatic is to minimise the static accumulation on plastic. Examples of antistatic include surfactants.

Lubricants Assist in processing and also help in stopping the sticking of plastic to metal parts during fabrication. Examples of lubricants include waxes, liquid paraffin. [8]

Rubbers Rubbers are excellent material for forming seals used to form closures such as bungs for vials or in similar applications such as gaskets in aerosol cans. Rubbers is used mainly for the construction of closure meant for vials transfusion fluid bottles dropping bottles and as washers in many other types of product. Rubber components may be made from either natural or synthetic sources.

Natural Rubbers Suitable for multiple use closure for injectable products as rubber reseals after multiple insertion of needle. It has certain degree of moisture and gas permeation.

Synthetic Rubber Have fewer additives and thus fewer extractable and tends to experience less sorption of product ingredient. Less suitable for repeated insertions of needle, because they tend to fragment or core pushing small particles of the rubber in to the product.

Table 5.2 Common packaging material used for different dosage forms

Dosage form	Administration route	Package material
Solid (tablet capsule powder)	Oral	Glass or plastic bottle Blister and strip packaging Sachet and pouches
Liquid	Oral	Glass or plastic bottle and cap Bottle with proper dropper Drums and jar Sachet and pouches
Liquid	Topical	Collapsible tube Aerosol spray
	Parenteral Ophthalmic	Glass ampoule Glass or plastic vial with applicator
Ointment cream	Topical Ophthalmic	Collapsible tube Glass or plastic jar Glass or plastic bottle Soft gelatin capsule

5.4.4   Common Packaging Techniques

Pharmaceutical packaging is one of the largest sectors of industry and is quickly becoming an integral part of the drug delivery system. Some of the common approaches for packaging of pharmaceuticals are discussed below (summarized in Table 5.2).

5.4.4.1   Blister Packaging

Strip packs and blister packs are well-known forms of pharmaceutical "unit dose" packaging that came into being in late 1920s and the early 1960s, respectively. Blister packaging is manufactured from a preformed plastic sheet in which blister cavities are formed under a vacuum which holds a product in place. Blisters can be of two types:

- Blisters having a cavity made up of thermoformed plastic, and the back of the blister is made of clear plastic/paper/aluminum foil.
- Blisters having a cavity and backing material both made from aluminum foil, which are formed by cold stretching [9, 10].

Infrared heaters or contact heaters are the main heating systems that are used to heat the feeding material web in thermo-cold forming blister. Application of heat can be done in different stages. Mostly two types of heating are done either direct or indirect. Direct heating is applied during forming stage, while indirect heating is applied at the indirect heat station with or without further heating. Plates or cylinders may serve as preheat stations. In some situations, the plates may be differentially heated. For cold-forming blister machine, the sequence involves the following steps:

- Heating the plastic material
- Formation of blister cavity by thermoforming or under vacuum
- Loading the blister with ingredient
- Placing the wrapping material over blister and sealing the package under heat

5.4.4.2 Strip Packaging

Strip pack is an alternative form for the unit dosage form. Mostly it is produced from paper, plastic or aluminum foil. The strip pack can be sealed by heat or pressure cold "self-adhesive" seal. A wide variety of materials can be used in making strip packs, from permeable sheet to those which incorporate a foil sheet of sufficient thickness that an individual hermetic seal is produced for each dosage unit. Compared to blister packaging, strip packaging requires less space. Moreover, the cost of the blister pack varies according to the material used [11].

Strip Packaging Process
Compared to blister packaging machines, strip packaging machines are quite simple and smaller. The common parts of the strip packaging unit are:

- Tray system
- Product insertion
- Heat sealing

Change in speed is possible. For example, with aspirin 325 mg, product speed of 250–300 per track is possible. Vacuum extraction is the process that is mainly used for powder, chip, and other material removal.

Those materials where the foil is used offer more protection because it ensures an excellent sealing, but caution is necessary while handling the foil so that over stretching of foil does not result in any perforation.

Aluminum/Aluminum Packaging Aluminum foil is a multilayered structure. It offers an excellent barrier to moisture. It is specifically designed for those pharmaceutical products which are heat-, moisture-, and light-sensitive and which cannot be packed suitably with barrier plastic films.

PVC/Aluminum This is suitable for packaging of ALU/PVC blister pack with option for Alu/Au foil blister packs. It is excellent for packing tablets or capsules.

OPA/PVC/ALU It provides almost for products that requires strict protection.

PVC/PVDC It is basically used for aluminum foil in blister packs.

PVC/PE This type provides the best barrier protection from moisture, oxygen, and light.

Strip packs with almost zero risk of reopening provide superior and excellent protection against moisture compared to other enclosable packs. The main difference between the strips and blisters is that strips are produced at lower speed and occupy more space compared to blisters which can be produced at high speed and require less space. So, in general, strips are more expensive compared to blisters.

Cellulose made or other films (single ply) are weak and cannot withhold external pressure, so most likely to break. To enhance the production speed, we have a variety of machines like the one refined machine handling 1-inch effervescent tablets offer high speeds of over 7000 items per minute faster than a simple blister machine.

5.4.4.3 Collapsible Tubes

The collapsible tube is extensively used in the pharmaceutical industry generally for viscous liquid or semisolid preparation mostly applied on the surface of the body-like skin and mucous membrane. Moisture and air are the two factors to which pharmaceutical products are quite sensitive. It may cause degradation of these products. Basically, tube is a cylindrically hollow piece with a round or oval piece.

Composition of Tube: Tubes can be made of a variety of materials but the most common ones are plastic and metals. Most tubes are designed to be dispensed with hand pressure, but some are used with a key at the base to roll them to help.

Plastic Tube
Polyethylene polymer is most commonly used to produce plastic tubes. Depending upon the barrier and printing properties, a variety of plastic films can be used. Different plastic films can be used for better barrier properties and printing. These are popular for cosmetic products, such as hand cream, and foodstuff. The plastic tubes retain their shape after each squeeze, such as a toothpaste. It can be highly decorated or made to use simply. Plastic tubes are produced by the extrusion method.

Aluminum Tubes
Aluminum tube caps and closure are generally threaded as a nozzle. The aluminum tube is hermetically sealed and is impervious to moisture and air. The aluminum tube blocks moisture and air and helps in product compatibility and quality for long-term use. Impact extrusion is the technique, through which these tubes are manufactured. The product can be squeezed out by finger pressure. It is also used for cosmetic pharmaceutical [12].

Pharmaceutical products that are used for skin and mucous membrane are mostly semisolid preparations. For these products, packaging metal tubes are most widely used. Metal tubes protect these preparations from the atmosphere as they may react with oxygen and cause damage to the product and may reduce its shelf life. The metal tubes especially the aluminum collapsible tube have collapsible or dead-fold properties which do not allow to keep the air inside the tube and which again reduce the chance of reaction of the product with the oxygen of the atmosphere. Lacquer coating is an additional coating in collapsible tubes, whose quality and uniformity are necessary for product compatibility and quality for long-term use and study [12].

At the present time in the pharmaceutical industry, pharmaceutical packaging technology is considered an important technique. When the product is ready, the packaging is the next step. As mentioned above pharmaceutical packaging is an important process as it gives a final and elegant look to the product which helps in patient compliance. Also, it helps in the physical protection of the products against extraneous factors like light, humidity, etc. Packaging also helps in product identification. As packaging can be sometimes a cost increasing factor for pharmaceutical companies, these pharmaceutical industries are now working extensively to increase their manufacturing and reduce the packaging costs by using different technologies, which give good quality packs, good sales, and economical results. The new modern marketing strategies and vast markets are the two important factors that have increased the demand for packaging products [13, 14].

References

1. Dean DA, Evans ER, Hall IH. Pharmaceutical packaging technology: CRC Press; 2005.
2. Monograph T. European Pharmacopoeia. European Directorate for the Quality of Medicine & Health Care of the Council of Europe (EDQM), edn. 2017;9:3104-5.
3. Amarji B, Kulkarni A, Deb PK, Maheshwari R, Tekade RK. Package Development of Pharmaceutical Products: Aspects of Packaging Materials Used for Pharmaceutical Products. Dosage Form Design Parameters: Elsevier; 2018. p. 521-52.
4. Mehta RM. Dispensing Pharmacy: Vallabh Prakashan; 2003.
5. Brunaugh AD, Smyth HD, Williams III RO. Nasal drug delivery. In Essential Pharmaceutics. Springer 2019; 183–187.
6. Jenke D. Extractable/leachable substances from plastic materials used as pharmaceutical product containers/devices. PDA Journal of Pharmaceutical Science and Technology. 2002;56(6):332-71.
7. Lachman L, Lieberman HA, Kanig JL. The theory and practice of industrial pharmacy: Lea & Febiger Philadelphia; 1976.
8. Allen L, Ansel HC. Ansel's pharmaceutical dosage forms and drug delivery systems: Lippincott Williams & Wilkins; 2013.
9. Pilchik R. Pharmaceutical blister packaging, Part I. Pharmaceutical technology. 2000;24(11):68-.
10. Remington JP. Remington: The science and practice of pharmacy: Lippincott Williams & Wilkins; 2006.
11. Piechocki JT. The History of Pharmaceutical Photostability Development. Pharmaceutical Photostability and Stabilization Technology. 2006;163:1.
12. Mallik J, Alam MF, Hossain SM, Rahaman M, Begum F, Das J. International Journal of Modern Pharmaceutical Research. 2018;7(1):29–37.
13. Chauhan AK, Patil V. Effect of packaging material on storage ability of mango milk powder and the quality of reconstituted mango milk drink. Powder Technology. 2013;239:86-93.
14. Ribeiro-Santos R, Andrade M, de Melo NR, Sanches-Silva A. Use of essential oils in active food packaging: Recent advances and future trends. Trends in food science & technology. 2017;61:132-40.

Chapter 6
Oral Emulsions

Saeed Ahmad Khan and Saddam Uddin

Abstract Drugs having low aqueous solubility can be delivered orally in the form of emulsion. For example, when drug is delivered through oily phase of emulsions, upon oral administration the oil droplet along with the active drug is absorbed through normal absorption mechanism of oil. Moreover, oil having therapeutic activity can also be administered in the form of emulsions. This chapter presents the pharmaceutical importance of oral emulsions, its types, and stability. Furthermore, the different factors to consider during formulation of stable emulsions, with particular emphasis on types of emulsifying agents, are discussed. Furthermore, theories of emulsification and quality control tests for emulsions are also discussed in this chapter.

Keywords Emulsifiers · Hydrophilic-lipophilic balance · Emulsion instability · Emulsion manufacturing

6.1 Introduction

Pharmaceutical emulsion is the heterogenous system composed of at least two immiscible liquid phases, one of which is dispersed in the form of small droplets, called *dispersed phase* or *internal phase*, in the other liquid phase, called *dispersion medium* or *continuous phase*, stabilized by *emulsifying agent(s)*. These two phases of the emulsions differ considerably from each other on the basis of their rheological properties [1]. Topical emulsion, such as creams exhibit pseudoplastic property and are more viscous as compared to emulsions that are administered orally.

S. A. Khan (✉) · S. Uddin
Department of Pharmacy, Kohat University of Science and Technology, Kohat, Pakistan
e-mail: saeekhan@kust.edu.pk

© The Author(s), under exclusive license to Springer Nature Switzerland AG 2022
S. A. Khan (ed.), *Essentials of Industrial Pharmacy*, AAPS Advances in the
Pharmaceutical Sciences Series 46, https://doi.org/10.1007/978-3-030-84977-1_6

Advantages Being a liquid dosage form, oral emulsions are favorable for patients who are unable to swallow solid dosage forms. The solubility and permeability of drugs can be increased when administered as emulsions [2]. Moreover, the bitter taste of drugs can be masked by dissolving the drug in oil component of an o/w emulsion, containing suitable sweetening/flavoring agent(s) in the dispersion medium [3]. Furthermore, oil having therapeutic effect, e.g., liquid paraffin, can be administered as small droplets through o/w emulsion. Besides, the skin irritancy of some drugs are decreased by formulating within an emulsion [4]. Emulsion is also used for *Total Parenteral Nutrition*.

Limitations Formulation of thermodynamically stable emulsions is a challenging task. Additionally, the manufacturing of pharmaceutical emulsions requires hectic procedures.

6.2 Types of Emulsions

Emulsions are essentially classified as simple emulsion and multiple emulsions, based on dispersed phase and dispersion medium. *Simple emulsions* are *oil in water (O/W) emulsions* or *water in oil (W/O) emulsions.* In O/W emulsions oil exists as internal phase and the dispersion medium is primarily water, e.g., milk. Contrarily, water in oil (W/O) emulsions contain water as dispersed phase and oil as dispersion medium, e.g., cold creams.

Multiple emulsions are formed by the dispersion of primary simple emulsion in another dispersion medium. Multiple emulsions are specially designed for [1] targeted drug delivery and [2] prolonging the release of drugs that have short half-life. Multiple emulsions may be either W/O emulsions being dispersed in water (W/O/W) or O/W emulsions being dispersed as globules in oil (O/W/O) as shown in Fig. 6.1.

Multiple emulsions are considerably under investigation from the previous few years. However, there is no commercial success yet. One of the main reasons is that they possibly revert back to their principal condition [5]. For example, improper storage of W/O/W emulsion can regress back to O/W emulsion.

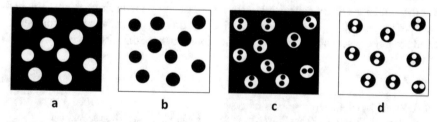

Fig. 6.1 Representation of different types of emulsion, W/O emulsion (**a**), O/W emulsion (**b**), O/W/O emulsion (**c**), and W/O/W emulsion (**d**). In image, gray color denotes oil phase, and white color represents aqueous phase

6.3   Theory of Emulsification

Emulsions are the heterogenous systems that are composed of liquid droplets distributed in another immiscible liquid. For instance, an oil in water dispersion will increase the surface area at interface between the two phases; the smaller the droplets, the higher will be the surface area [6]. Hence, an interfacial tension exists (higher interfacial Gibbs free energy) at the interface between the two phases, that make the system unstable, as shown in Eq. 6.1:

$$\Delta G = \gamma_{o/w} \Delta A \tag{6.1}$$

where ΔG represents the Gibbs free energy at the interface, $\gamma_{o/w}$ is the surface tension between oil phase and water phase at the interface, and ΔA is the change in surface area after dispersion. The interfacial Gibbs free energy gets reduced when the droplets undergo *coalescence*, i.e., formation of large single droplet, having less exposed surface area per unit volume [6].

In order to stabilize the system and prevent phase separation, *emulsifying agent(s)* (also called *emulsifier* or *surfactant*) is added. There are several theories about how emulsifier(s) stabilizes an emulsion. The most common theories are oriented wedge theory, interfacial tension theory, viscosity theory, and repulsion theory.

Oriented Wedge Theory This theory proposes the formation of monomolecular layer of emulsifier curved around the dispersed droplets. Emulsifiers and amphiphiles will be preferentially dissolved in water or oil, depending on the predominant hydrophilic or hydrophobic portion [5]. For instance, an emulsifier with predominant hydrophilic part would be preferentially dissolved in water and would favor formation of O/W emulsion and vice versa.

Interfacial Tension Theory This theory says that emulsifier lowers the interfacial tension between the two phases, thus resulting into a stable emulsion [1, 6].

Viscosity Theory This theory mentions that emulsifiers increase viscosity of the dispersion medium, thus maintaining a stable dispersion of globules.

Repulsion Theory This theory proposes that emulsifiers create a repulsive film over the dispersed droplets, as a result the globules remain suspended in the dispersion medium.

From the above discussion, it is understood that emulsifier is an important component of emulsion formulation.

6.4   Formulation Considerations of Emulsions

For formulation of stable pharmaceutical emulsion, the choice of excipients depends on factors such as the desired type of emulsion, the desired route of administration, volume of internal phase, size of the globules, and the desired consistency of emulsion. All these aspects are individually explained below.

6.4.1   Emulsion Type

Emulsion (W/O or O/W) might vary on the basis of their clinical applications. Emulsions that are administered orally or by IV route are primarily O/W type, while topically applied emulsions (creams) are W/O or O/W type. Generally, O/W creams are used for local effect of water-soluble drug on the skin, in conditions like inflammation or infection [5]. They are non-greasy and smoothly applied to the skin and may be washed from the skin easily, while W/O emulsions are greasy in nature and hydrate the skin, e.g., moisturizing creams are mostly W/O emulsions.

6.4.2   Volume of the Internal Phase

The volume of internal phase has substantial influence on the stability of emulsions. For O/W emulsions, the ratio of internal-to-external phase can be 1:1; however, for W/O emulsion the maximum concentration of internal phase is 30–40% (i.e., 1:1.3), to formulate stable emulsion. Since, higher concentrations will result in phase inversion.

6.4.3   Droplet Size

The droplet size of internal phase affects the stability of emulsions. Smaller droplets are more challenging to be [8] stabilized, while bigger droplets exhibit creaming and sedimentation. Therefore, to ensure the stability of emulsion, size of the droplet is kept optimum.

6.4.4   Viscosity

The viscosity of internal phase and dispersion medium affects the type of emulsion produced. The chances of coalescence of dispersed droplets are reduced by increasing the viscosity of dispersion medium [5]. Moreover, stability of emulsion is also affected by variation in the viscosity of final formulations because it controls the speed of creaming and sedimentation. As a general principle, when the viscosity of one phase is increased, the chances of that phase being the external phase are also increased [1].

The topically applied emulsion (creams) have higher viscosity compared to oral emulsions and parenteral emulsions. Creams, being intended for topical applications have greater viscosity, which promotes greater retention and spreading of cream on the skin.

6.4.5 Preservation

For preservation of emulsions, proper and sufficient amount of antimicrobial agent is required. Preservation of emulsion is a challenging task, because the antimicrobial agent may partition into the oil phase and hence the concentration of antimicrobial agent in aqueous phase might deplete below the minimum inhibitory concentration (MIC) and ultimately become ineffective [9]. In such cases, slightly higher concentration of antimicrobial agent is used. Common antimicrobial agents for emulsion are benzalkonium chloride (also used as emulsifier), methyl and propyl parahydroxybenzoic acid, benzoic acid, etc.

Besides, antioxidant is also needed in order to preserve emulsion for longer time. In emulsion, other than therapeutic agent, both oil and water phases are also prone to oxidation. Therefore, lipophilic antioxidants, such as butylated hydroxyanisole, butylated hydroxytoluene, propyl gallate, etc., are added to oil phase, while hydrophilic antioxidants, such as sodium metabisulfite, sodium sulfite, etc., are added to water phase.

6.4.6 Type and Concentration of Emulsifying Agents

Emulsifying agents are the substances that reduce the interfacial tension between oil phase and aqueous phase, thus resulting in the formation of stable emulsions. All pharmaceutical emulsions require inclusion of emulsifier, to ensure the stability of emulsions. The type of emulsifier selection is dependent upon the nature of desired emulsion.

Surface active agents typically possess both lipophilic and hydrophilic groups. The contribution of these groups determines that the surfactant is either lipophilic or lipophobic (hydrophilic) [5], which ultimately dictates the type of emulsion formed. For instance, hydrophilic emulsifying agents form O/W emulsion, while lipophilic emulsifiers yield W/O emulsion [10]. This contribution is commonly referred as *hydrophilic-lipophilic balance* (HLB). HLB is a ratio scale that gives a number to each surfactant on the basis of the individual groups. HLB value is a useful tool for selection of surface-active agent. HLB values typically ranges from 1 to 20,

Table 6.1 HLB guide for selection of surfactant

Function of surfactant	HLB value
Antifoaming agents	1–3
Emulsifiers for W/O emulsion	3–6
Wetting agents	7–9
Detergents	13–16
Emulsifiers for O/W emulsion	8–18
Solubilizers	15–20

Adopted from Ref. [11]

however, some emulsifiers have shown higher HLB values, experimentally, e.g., Poloxamer 407 and Poloxamer 188 has HLB value 22 and 29 respectively. As the HLB value increases, the hydrophilic nature of emulsifying agents also increases. The HLB value of emulsifying agents along with primary functions are given in Table 6.1.

For selection of emulsifying agent, HLB requirement of dispersed phase of emulsion is an important consideration. Information about the required HLB to form a stable O/W or W/O emulsion is available in literature. For instance, formulation of stable O/W emulsion, in which cottonseed oil is used as droplets [12] of internal phase requires emulsifier(s) with HLB value of 10. Whereas, W/O emulsion having cottonseed oil as external phase needs emulsifier(s) having HLB value of 5. If a mixture of two emulsifiers is employed, the fraction of each emulsifier can be determined by using simple weighted-average equation, where x is the fraction of emulsifier 1 and $1 - x$ is the amount of emulsifier 2:

$$HLB_{required} = x\,HLB_{emulsifier1} + (1-x)\,HLB_{emulsifier2} \tag{6.2}$$

For instance, a mixture of Span 60 (HLB 4.7) and Tween 80 (HLB 15.0) would need a ratio of 0.49:0.52 for preparation of cottonseed oil O/W emulsion, and a ratio of 0.97:0.03 would be required for W/O emulsion of cottonseed oil.

If the HLB requirement is unknown, then a large number of emulsion formulations are prepared using numbers of emulsifiers, thus giving a wide range of HLB values by using weighted-mean approach. The most stable formulation is selected for formulation development.

6.4.6.1 Classification of Surface-Active Agents

One of the most common classifications of emulsifiers is based on the charge they possess.

Anionic Surfactants These emulsifiers produce negatively charged ion when dissociated. They are inexpensive but comparatively unsafe if taken in large amount, therefore, primarily used in topically applied products. Examples of anionic emulsifiers are summarized in Table 6.2.

Table 6.2 Examples of anionic surfactants

Examples	Emulsion type	Application route
Monovalent fatty acid salt: e.g., sodium oleate, sodium stearate, ammonium oleate, etc.	O/W	Topical, oral
$H_3C(CH_2)_6$ $(CH_2)_6COONa$ Chemical structure of sodium oleate		
Divalent salts of fatty acids: e.g., calcium stearate	W/O	Topical, oral
$[CH_3(CH_2)_{15}CH_2\overset{O}{\overset{\|}{C}}O]_2 Ca^{+2}$ Chemical structure of calcium stearate		
Alkyl sulfates: sodium lauryl sulphate and triethanolamine lauryl sulphate	O/W	Topical, oral
H_3C ... $O-Na^+$ Chemical structure of sodium lauryl sulfate		

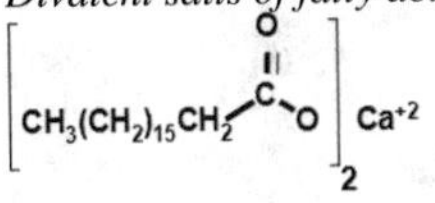

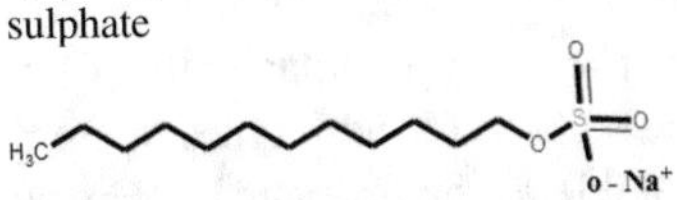

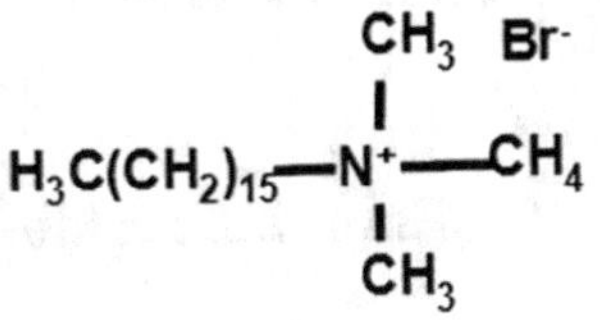

Fig. 6.2 Chemical structure of hexadecyl trimethylammonium bromide (cetrimide)

$$H_3C(CH_2)_{15}-\overset{\overset{CH_3}{\|}}{\underset{\underset{CH_3}{\|}}{N^+}}-CH_4 \quad Br^-$$

Cationic Surfactants These emulsifiers upon dissociation produce positively charged species having emulsifying property. Cationic emulsifiers possess antimicrobial property and hence are primarily used as preservatives in topical formulation. Nevertheless, in combination with nonionic surfactant having low HLB value, [13] these emulsifiers may be used for formation of O/W emulsions.

Examples include cetrimide, which is a mixture of triethylammonium bromide, with minute concentration of dodecyl trimethylammonium bromide and hexadecyl trimethylammonium bromide. The general structure is shown in Fig. 6.2.

Amphoteric Surfactants Surface-active agents contain both positive and negative charged species depending upon pH value. At low pH they liberate cation while at high pH they generate anion.

The emulsifying property of these surfactants is decreased when the pH reaches the isoelectric point. The most commonly used amphoteric surfactants are phospholipids, also called lecithin (shown in Fig. 6.3), which is generally isolated from soybean or egg yolk. Lecithins are usually used in combinations with other emulsifiers.

Fig. 6.3 Chemical
structure of lecithin

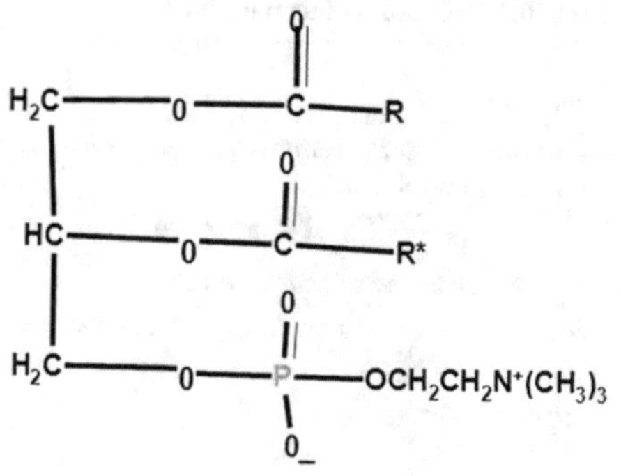

Nonionic Surfactants Nonionic emulsifiers are the most popular category of emulsifiers. Depending on the HLB value, they can be used for both types of emulsions. Being uncharged, nonionic surfactants are stable to the changes in pH or to the presence of electrolytes. Lipophilic part of these emulsifiers is made from fatty acid or fatty alcohol, while hydrophilic part is either -OH or ethylene glycol moieties. Examples are summarized in Table 6.3.

6.5   Emulsion Instability

In development of formulation for emulsion, the primary goal is to ensure that the globules retain their initial integrity and remain uniformly dispersed in dispersion medium. However, it is not always possible; the emulsion may sometimes exhibit reversible or irreversible instability [13]. Some of the common phenomena that contribute to emulsion instability are cracking (irreversible instability), flocculation, creaming, and phase inversion, as shown in Fig. 6.4.

6.5.1   Cracking

Cracking involves the complete coalescence of the internal phase, which eventually leads to phase separation. Cracking is an irreversible instability and may be due to the following factors.

Improper or Insufficient Emulsifier Using incorrect emulsifier(s) or insufficient emulsifier(s) leads to the formation of interfacial layer that has deficient mechanical strength; thus coalescence of droplets cannot be prevented.

Complexation of Formulation Components with Emulsifier It is necessary that the drug or excipients used are inert, having no interaction with emulsifier, so that

Table 6.3 Example of nonionic surfactants

Class of nonionic surfactant	Examples	HLB value
Sorbian esters (Span series)		
	Sorbitan monolaurate (Span 20®) R_1: $C_{11}H_{23}$ R_2: H R_3: H	8.6
	Sorbitan monopalmitate (Span 40®) R_1: $C_{15}H_{31}$ R_2: H R_3: H	6.7
	Sorbitan monostearate (Span 60®) R_1: $C_{17}H_{35}$ R_2: H R_3: H	4.7
	Sorbitan tristearate (Span 65®) R_1: $C_{17}H_{35}$ R_2: $C_{17}H_{35}$ R_3: $C_{17}H_{35}$	2.1
Polyoxyethylene fatty acid derivatives of the sorbitan esters (Tween series)		
	Polyoxyethylene sorbitan monolaurate (Tween 20) R: $C_{11}H_{23}$	16.7
	Polyoxyethylene sorbitan monolaurate (Tween 40) R: $C_{15}H_{31}$	15.6
	Polyoxyethylene sorbitan monostearate (Tween 60) R: $C_{17}H_{35}$	14.9
	Polyoxyethylene sorbitan monooleate (Tween 80) R: C7H14CH=CHC8H17	15.0
Polyoxyethylene alkyl ethers (Brij series)		
$CH_3(CH_2)_x(OCH_2CH_2)_yOH$	Polyoxyethylene [4] lauryl ether (Brij 30®) x:11 y: 4	9.7
	Polyoxyethylene (20) cetyl ether (Brij 58®) x:15 y: 20	15.7
	Polyoxyethylene [2] oleyl ether (Brij 52®) x:15 y: 2	4.9
Poly(ethylene oxide) poly(propylene oxide)-poly(ethylene oxide) copolymers (Poloxamer series)		
	Poloxamer 181 (Pluronic® L61) a: 2 b: 31	3
	Poloxamer 407 (Pluronic® F127) a: 100 b: 65	22
	Poloxamer 188 (Pluronic® F68) a: 77 b: 29	29

the interfacial layer is not affected. For instance, if drug, preservatives, or any other excipient possess charge opposite to that exhibited by the interfacial film, complexation may occur, that eventually destructs the interfacial film.

Temperature Emulsions are stable at optimum temperature. Variations in the temperature generally affect the stability of emulsion, because an increase in temperature may coagulate certain types of emulsifying agent (e.g., proteins). Similarly lowering the temperature, i.e., freezing of the aqueous phase produces ice crystal that exerts unusual pressure on the oil globules.

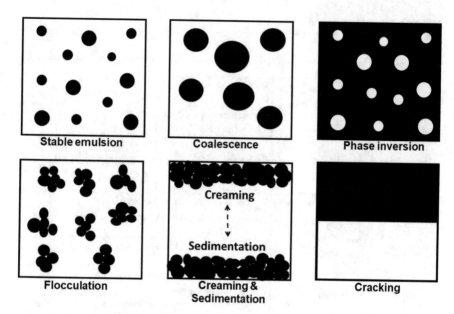

Fig. 6.4 Schematic representation of different types of emulsion instability

Microbial Spoilage Microorganisms may metabolize the surface-active agents, thus destroying the interfacial film and ultimately destabilizing the emulsion.

6.5.2   Creaming and Sedimentation

The droplets of internal phase may form loose aggregates due to the presence of weak van der Waals forces. This phenomenon is called *flocculation*. The integrity of the droplets remains intact, and these weak forces keep the droplets at a certain distance from each other. The globules can be easily redispersed upon slight shaking [7].

However, there is possibility that upon close existence the droplets may undergo coalescence [8], if the mechanical property of interfacial is compromised. Globules of the discontinuous medium after coalescence tend to accumulate at the surface or settle at the bottom of emulsion, called *creaming* or *sedimentation*, respectively. It occurs because of the difference in the densities of two phases. The speed of creaming or sedimentation is governed by the Stokes law:

$$V = d^2 \left(\rho_i - \rho_m \right) g / \eta$$

(6.3)

where V = creaming or sedimentation rate, d = diameter of the internal phase droplet, ρ_i = density of internal phase, ρ_m = density of dispersion medium, g = gravitational force, and η = viscosity of dispersion medium.

The Stokes equation elaborates that if density of dispersed phase is lower than dispersion medium, as the case in O/W emulsion, the value of velocity is negative, which means the droplets move upward (creaming) and vice versa when internal phase is denser than dispersion medium. Moreover, the equation also shows that creaming and sedimentation can be prevented by lowering the difference in the densities of two phases or enhancing the viscosity of the formulation.

6.5.3 Phase Inversion

It represents the conversion of one emulsion into another, i.e., W/O emulsion to O/W emulsion and vice versa. Phase inversion typically occurs when the ratio of volume of dispersed phase to dispersion medium exceeds the required limit (i.e., 60:40 for O/W emulsion and 40:60 for W/O emulsion) [14]. Phase inversion is also caused by the addition of electrolyte or by temperature change. It may be controlled by choosing appropriate and sufficient amount of emulsifying agent.

6.6 Excipients Used in Emulsion

Like other pharmaceutical formulations, emulsion also requires proper excipients in appropriate concentration, to ensure the physicochemical stability of emulsion and to make the product pleasing to the patient. For this purpose, the number of excipients such as emulsifying agent, buffers, preservatives, etc. is added and these are discussed below in the Table 6.4.

6.7 Manufacturing of Emulsions

Manufacturing of emulsion is a multistep process, involving the following steps:

Preparation of oil phase and water phase The first step involves the dissolution of oil-soluble components in oil phase and water-soluble components are [15] dissolved in water phase separately.

Emulsification Both the phases are mixed to form crude emulsion. This step typically requires turbulent mixing in order to ensure proper dispersion of internal phase in dispersion medium.

Homogenization Homogenization is required in order to reduce droplet size of the internal phase. Homogenization at laboratory scale is carried out by small kitchen mixers or mortar and pestle, while at industrial scale it employs mechanical stirrers, homogenizers, ultrasonifier, and colloidal mill.

Table 6.4 Common excipients used in oral emulsions

Excipients	Examples
Water phase	Purified water (USP)
Oil phase	*Vegetable oils:* Cottonseed oil, arachis oil, almond oil
Emulsifying agents	*Anionic:* Iodium oleate, calcium stearate *Cationic:* Cetrimide *Amphoteric:* Phospholipids *Nonionic:* Sorbian esters (Span series), polyoxyethylene fatty acid derivatives of the sorbitan esters (Tween series)
Viscosity builders	*Cellulose derivatives:* Sodium carboxymethylcellulose, hydroxypropyl methylcellulose, etc. *Natural gums:* Acacia, Tragacanth, xanthan gum, guar gum, etc. *Natural polymers:* Chitosan, agar, alginate, etc. *Synthetic polymers:* Carbopol, polyvinylpyrrolidone, polyvinyl alcohol, etc.
Buffers	Citric acid/sodium citrate Acetic acid/sodium acetate
Antimicrobial agents	Methyl and propyl parahydroxybenzoic acid, benzoic acid, etc.
Antioxidants	*Hydrophilic antioxidants:* Sodium sulfite, sodium metabisulfite, ascorbic acid, etc. *Hydrophobic antioxidants:* Ascorbyl palmitate, tocopherol, etc.
Sweetening agents	Dextrose, fructose, glucose, aspartame, maltose mannitol, saccharin, sorbitol, etc. (for oral emulsions only)
Other excipients	Specific colorants and flavors may also be used (for oral emulsions only)

References

1. Zhu Y, Gao H, Liu W, Zou L, McClements DJ. A review of the rheological properties of dilute and concentrated food emulsions. Journal of Texture Studies. 2020;51(1):45–55.
2. Sutradhar KB, Amin ML. Nanoemulsions: increasing possibilities in drug delivery. European Journal of Nanomedicine. 2013;5(2):97–110.
3. Vandamme T. Reverse water-in-oil microemulsions as taste masking of bitter drugs intended for oral administration. STP pharma practice. 2003;13(1):8–12.
4. Cunningham CT, Seidling JR, Kroll LM, Mundschau SA. Stable emulsion for prevention of skin irritation and articles using same. Google Patents; 2017. WO2014002037A3
5. Akbari S, Nour AH. Emulsion types, stability mechanisms and rheology: A review. International Journal of Innovative Research and Scientific Studies 2018;1(1):14–21.
6. Friberg SE, Quencer LG, Hilton ML. Theory of emulsions in Pharmaceutical dosage forms: Dispersed systems volume 1. 2020; CRC Press, Boca Raton, Florida; pp. 53–90.
7. Dickinson E. Strategies to control and inhibit the flocculation of protein-stabilized oil-in-water emulsions. Food hydrocolloids. 2019;96:209–23.
8. Yao J, Lin F, Kim HS, Park J. The effect of oil viscosity on droplet generation rate and droplet size in a T-junction microfluidic droplet generator. Micromachines (Basel). 2019;10(12):808.
9. El-Guendouz S, Aazza S, Dandlen SA, Majdoub N, Lyoussi B, Raposo S, et al. Antioxidant activity of Thyme waste extract in O/W emulsions. Antioxidants (Basel). 2019;8(8):243.
10. Zakinyan AR, Kulgina LM, Zakinyan AA, Turkin SD. Electrical Conductivity of Field-Structured Emulsions. Fluids, 2020;5(2):74.
11. Brunaugh AD, Smyth HD, Williams III RO. Essential Pharmaceutics: Springer Nature Switzerland; 2019: 111–121.

12. Wei Z, Huang Q. Edible Pickering emulsions stabilized by ovotransferrin–gum arabic particles. Food Hydrocolloids. 2019;89:590–601.
13. Matubayasi N. Surface tension and related thermodynamic quantities of aqueous electrolyte solutions: CRC Press, Boca Raton, Florida. 2013:48
14. Geng T, Qiu Z, Zhao C, Zhang L, Zhao X, et al. Rheological study on the invert emulsion fluids with organoclay at high aged temperatures. Colloid and Surfaces. 2019;573:211–221.
15. Khan BA, Akhtar N, Khan HMS, Waseem K, Mahmood T, Rasul A, et al. Basics of pharmaceutical emulsions: A review. African Journal of Pharmacy and Pharmacology 2011;5(25):2715–2725.

Chapter 7
Oral Suspensions

Saeed Ahmad Khan, Abdul Baseer, Salar Khan, and Mehmood Hussain

Abstract The use of suspension dosage form has a long history for poorly soluble drugs. Development of stable suspension of the drug product has been a challenge on many fronts. In this chapter some of the important aspects of stable suspension formulation are discussed. Particularly, properties related to drug delivery and suspension stability are explained in detail. Suspension with regard to theoretical consideration including flocculation, viscosity, particle settling, zeta potential, Ostwald ripening, and particle aggregation is explained. Discussion relevant to excipients is also included. Moreover, methods of particle size reduction are introduced, including those employed for nanosuspension.

Keywords Electric double layer · DLVO theory · Flocculation · Coagulation · Manufacturing of suspension

7.1 Introduction

The dispersion system in which the drug (dispersed phase) is dispersed as finely divided insoluble particles in the vehicle (dispersion medium) is termed as pharmaceutical suspension [1].

The dispersed phase tends to separate on storage; therefore, it is important to control the process of separation, usually with the help of suspending agent(s) or surfactant(s). A stable suspension is the one that ensures an accurate dose after

S. A. Khan (✉) · M. Hussain
Department of Pharmacy, Kohat University of Science and Technology, Kohat, Pakistan
e-mail: saeekhan@kust.edu.pk

A. Baseer
Department of Pharmacy, Abasyn University, Peshawar, Pakistan

S. Khan
Department of Pharmacy, Abdul Wali Khan University Mardan, Mardan, Pakistan

shaking. Ideally, pharmaceutical suspension should have the following properties [2]:

- It should exhibit low sedimentation rate.
- It should be easily redispersible on gentle shaking.
- It should be easily dispensed from container.
- It should be acceptable for patients in terms of taste, smell (in case of oral suspensions), and administration.

Advantages Pharmaceutical suspension is used for a variety of pharmaceutical uses, for instance, drugs having low solubility, e.g., prednisolone, are formulated as suspensions [3]. Furthermore, the degradation of some drugs, e.g., oxytetracycline, might be prevented when they are formulated as suspension [1]. Suspension can also be used to mask the bitter taste of drugs, e.g., chloramphenicol palmitate suspension. Moreover, suspension can be helpful in controlling the release of drugs, e.g., procaine penicillin. Besides, in some cases, suspension dosage form is the only option available for administration of drugs, e.g., some vaccines like cholera vaccine [4].

Disadvantages Pharmaceutical suspensions may exhibit physical instability, e.g., sedimentation and cracking during shelf life. Special formulation measures are required to minimize instability and to ensure uniform dosing. Moreover, formulation of aesthetically good suspension is very challenging. Besides, suspensions require careful handling during transport [4].

7.2   Dispersed Particles in a Medium

The particles when dispersed in a medium exhibit electrical charges either due to ionization of functional groups, e.g., COOH and NH_2, or by adsorbing ions to its surface. Once the surface of dispersed particles gets a charge, a phenomenon called *electrical double layer* takes place [5], as shown in Fig. 7.1. The charged particle attracts counterions that form an immovable layer around the particle and is called the stationary layer or *stern layer*. However, the surface charge of particle is not completely neutralized. This residual surface charge attracts more counterions from its surrounded medium so that a second layer of counterions is developed around the particle. This layer is farther away from the particle surface. The attractive forces of the particle charged surface get weaker with the distance. Therefore, this second layer is less ordered and movable and is called diffuse layer. Both anions and cations are present in diffuse layer, but ions that are of opposite charge to the surface predominate. There exists an imaginary boundary within the diffuse layer, and inside this boundary the particle acts as a single entity, called *slipping plane*. The potential at slipping plane is said to be *zeta potential* [6].

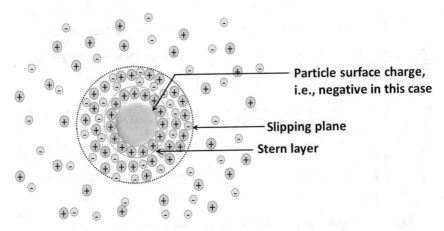

Fig. 7.1 Diffuse double layer model of a positively charged surface in an aqueous medium

The magnitude of zeta potential shows the degree of electrostatic repulsion between two nearest particles. However, along with the electrostatic repulsive forces, van der Waals forces of attraction also exist. The well-known Derjaguin-Landau-Verwey-Overbeek theory (*DLVO theory*) suggests that the van der Waals attractive forces and electrostatic repulsive forces can be superimposed at each interacting distance between adjacent particles. So the overall energy of interaction between particles (Vt) is the sum of the energies of attraction (Va) and energies of repulsion (Vr) [7].

$$V_t = V_a + V_r \tag{7.1}$$

As shown in Fig. 7.2, three main regions may be observed in the total energy of interaction:

The primary minimum: This is a region of high degree of instability of formulation. In this region, particles exhibit strong attraction and tend to exhibit coagulation (irreversible aggregates).

The primary maximum: In this region, the repulsion interaction between particles predominates and particles tend to repel each other. The magnitude of repulsion is dependent on adjacent particles' zeta potential.

The secondary minimum: in this region, attractive forces predominate over repulsive forces; however, as compared to primary minimum, the magnitude of attraction is less. Consequently, the particles undergo *flocculation*, i.e., the formation of loose aggregates (floccules), as shown in Fig. 7.3. The sediment formed in flocculated suspension can be easily redispersed by gentle shaking. This phenomenon is sometimes exploited for *controlled flocculation* [8].

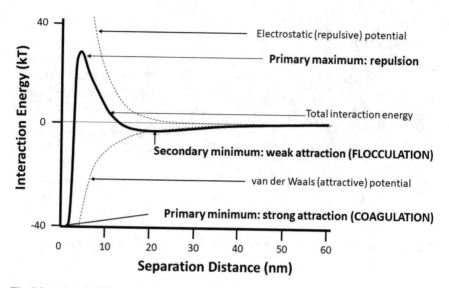

Fig. 7.2 Diagrammatic representation of the overall interactive energy between two particles and their distance of separation

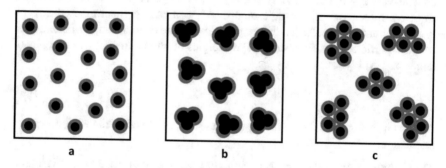

Fig. 7.3 Schematic representation of stable suspension (**a**), coagulation (**b**), and flocculation (**c**)

7.3 Formulation Considerations

The excipients used in formulation of oral suspensions are summarized in Table 7.1. The following section discusses various factors that need to be considered during formulation of stable pharmaceutical suspension.

Table 7.1 Common excipients used in pharmaceutical suspensions

Excipients	Examples
Vehicle	Purified water (USP) Glycerin, polyethylene glycol, propylene glycol, and alcohol
Buffers	Citric acid/sodium citrate Acetic acid/sodium acetate
Suspending agents/ viscosity builders	*Cellulose derivatives:* Sodium carboxymethylcellulose, hydroxypropyl methylcellulose, etc. *Natural gums:* Acacia, tragacanth, guar gum, xanthan gum, etc. *Natural polymers:* Chitosan, agar, alginate, etc. *Synthetic polymers:* Polyvinylpyrrolidone, carbopol, polyvinyl alcohol, etc.
Flocculating agents	*Electrolytes:* Sodium chloride, aluminum chloride, potassium dihydrogen phosphate, etc. *Surfactants:* Docusate sodium, sodium lauryl sulfate, sorbitan monolaurate, etc. *Hydrophilic polymer:* Tragacanth, sodium carboxymethylcellulose, etc.
Wetting agents	*Surfactants:* Phosphatidyl choline; sodium dodecyl sulfate; polysorbate (Tween®) 20, 40, 60, and 80; sorbitan esters (Span®) 40, 60, and 80; cetrimide; etc. *Hydrophilic polymers:* Gelatin, acacia, cellulose derivatives, etc.
Antimicrobial agents	Benzalkonium chloride (also used as surfactant), methyl and propyl parahydroxybenzoic acid, benzoic acid, etc.
Antioxidants	Ethylenediamine tetraacetic acid (EDTA), sodium metabisulfite, sodium sulfite, ascorbic acid, etc.
Sweetening agents (in case of oral suspension)	Dextrose, fructose, glucose, aspartame, maltose, mannitol, saccharin, sorbitol etc.
Other excipients	Specific colorants and flavors may also be used

7.3.1 Chemical Incompatibility

The compatibility of all the ingredients in suspension, i.e., buffers, drugs, and suspending agents should be thoroughly investigated. For instance, carboxymethyl cellulose (CMC) may interact with cationic excipients, e.g., benzalkonium chloride, and get precipitated. Therefore, the use of CMC should be avoided in the presence of cationic excipients [9].

7.3.2 Particle Size

The physical stability of a suspensions is directly dependent on the particle size of dispersed phase. According to Stokes equation, the rate of sedimentation (V) is directly proportional to particle size (d) and the difference of particle-vehicle densities ($\rho_s - \rho_o$) and inversely proportional to viscosity of the vehicle (η):

$$V = d^2 \frac{\left(\rho_s - \rho_o\right)g}{18\eta}$$

$$(7.2)$$

where:

V = sedimentation rate
d = particle diameter
ρ_s = particle density
ρ_o = vehicle density
g = gravitational force
η = vehicle viscosity

Dissolution of drug can also be affected by particle size. Reduction in particle size increases dissolution rate of drug; hence the bioavailability of drug may be increased.

Moreover, the particle size also depends on the intended route of application. For instance, too large particle size imparts gritty texture to the pharmaceutical suspension that can be unfavorable for ophthalmic and topical application. For this reason, particles size in ophthalmic and topical suspensions is typically around 1 µm and 10–100 µm, respectively [10].

An ideal suspension has a constant average particle size throughout the shelf life. However, sometimes a slight increase in storage temperature enables the smaller drug particles to dissolve. Recrystallization of dissolved drug occurs on the larger particle surfaces, thus increasing the average particle size. This phenomenon is called *crystal growth* or *Ostwald ripening* [11]. Inclusion of hydrophilic polymers in formulation can prevent crystal growth by adsorbing on to the surface of dispersed drug particles.

7.3.3　Viscosity

Viscosity is the property of a fluid to resist the flow (shear deformation). In technical terms, viscosity (η) is shear stress divided by shear rate.

Shear stress is the force experienced by a cross section of fluid to move the fluid layers past each other. *Shear rate* is the rate at which the fluid layers move past each other.

The fluids where shear rate linearly increases with shear stress, i.e., viscosity of fluid remains constant, these are called *Newtonian* fluids, e.g., water and honey. However, in reality, most of the suspensions exhibit *non-Newtonian* flow, and the fluid shows a nonlinear relationship between shear stress and shear rate, i.e., the viscosity of fluid varies with shear rate. Non-Newtonian fluids are called *pseudoplastic* or *shear-thinning* fluids when with increase in shear rate the viscosity of fluid decreases. However, when fluid viscosity increases with increasing shear rate, the fluid is termed as *shear-thickening* or *dilatant* fluid. In some fluids, a certain

Fig. 7.4 Graphical
representation of flow
property of various types
of fluids

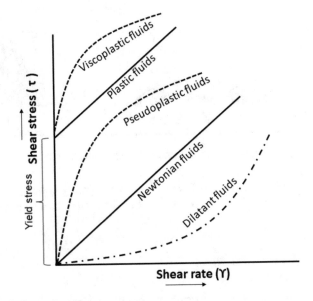

stress level, called *yield stress* or *yield value*, is required to initiate flow; these fluids if after the yield stress exhibit Newtonian flow are called *plastic* fluids or *Bingham fluids* [12]. And if the fluid exhibits shear-thinning behavior after yield stress, then the fluids are termed as *viscoplastic fluids*. The flow curves of different type of fluids are given in Fig. 7.4.

As mentioned earlier the rate of sedimentation is inversely proportional to the viscosity. The physical stability of suspension can be improved by increasing the viscosity. Adding a suspending agent is one of the most common methods of increasing the viscosity, although too high viscosity is undesirable due to poor pourability and dosing.

7.3.4   Wetting

For preparation of stable suspension, the solid dispersed particles should be completely wetted by the solvent. Since water is commonly used as vehicle for pharmaceutical suspension, therefore particles that are hydrophilic in nature are easily wetted by water. Contrarily, in hydrophobic materials water does not penetrate into the inter-particular spaces of powder. Therefore, wetting agent is added to improve wetting of the hydrophobic powder. The wetting property of material can be assessed by measuring the *contact angle*, which is the angle a liquid creates on a solid surface, as shown in Fig. 7.5.

Surfactants are frequently used as wetting agents in pharmaceutical suspensions. Surfactants act as wetting agent by decreasing the solid-liquid interfacial tension.

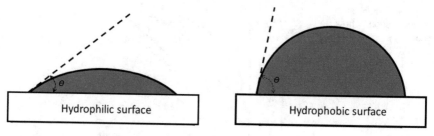

Fig. 7.5 Contact angle at hydrophilic and hydrophobic surfaces

Surfactants with HLB value 7–9 act as wetting agents [5]. Besides this, hydrophilic polymers, e.g., carboxymethyl cellulose, etc., are also used as wetting agent.

7.3.5 Mixing

Homogeneous dispersion of drug in vehicle requires efficient mixing. The mixers employed in manufacturing of pharmaceutical suspension must have the capacity to mix viscous material. Mixing time must be critically validated, since too much shearing results in accumulation of air in the bulk liquid.

7.3.6 Size Reduction and Homogenization

As discussed earlier, particle size can affect the sedimentation rate and redispersion of the sedimented particles. For small scale, the particle size can be reduced by pestle and mortar. For large-scale manufacturing, homogenizers, such as *colloid mill, piston gap homogenizer*, etc., described in Chap. 3 Sect. 7.2, are employed. Typically, homogenization is the last step in manufacturing of almost all types of pharmaceutical suspensions [13].

7.3.7 Flocculation

Flocculation is a phenomenon whereby small suspended particles form loose aggregates (floccules) that can be easily separated by gentle shaking. Flocculation of particles occurs when repulsive electrical forces in a dispersed system become low to the extent (i.e., secondary minimum) that the suspended particles can form floccules. The floccules settle and produce sediment which is less dense and easier to redisperse on gentle shaking than a sediment produced by deflocculated suspension. The major difference between flocculated and deflocculated suspension is given in Table 7.2. It should be noted that too much reduction in repulsive forces may result

Table 7.2 Physical attributes of flocculated and deflocculated suspension

Deflocculated suspension	Flocculated suspension
Particles are suspended as separate entities	Particles are suspended as loose aggregates
Slow rate of sedimentation	Fast rate of sedimentation
The sediment is hard to be redispersed	The sediment is easily redispersible
Due to uniform dispersion of relatively smaller particles, the suspension has a pleasing appearance	Due to dispersed floccules, the suspension has visible distinct phases

into irreversible aggregates of particles (coagulation). With the use of *flocculating agents* (e.g., electrolytes, polymers, surfactants, etc.), the flocculation process can be adequately controlled, called *controlled flocculation* [8].

7.3.8   Preservation

Suspensions for oral administration are multidose products; therefore, preservation against microorganisms is necessary. The limits regarding number and type of microorganisms in oral solutions and suspensions are specified in various pharmacopoeias. For instance, the European Pharmacopoeia specifies that oral products should be free from *E. coli;* moreover, the number of aerobic bacteria and fungi per gram or milliliter should be less than 1000 and 100, respectively.

Additionally, oral suspensions also require preservation against oxidation to enhance the chemical stability of drug [14]. Antioxidants are compounds (e.g., sodium sulfite, ascorbic acid, sodium metabisulfite, etc.) that get oxidized faster than the drug, hence preventing oxidation of drug. Chelating agents, e.g., ethylenediamine tetraacetic acid (EDTA), citric acid, etc., also act as antioxidant by forming complexes with heavy metal ions that are involved in oxidative degradation of drugs.

7.4   Industrial-Scale Manufacturing of Suspension

Pharmaceutical suspension is usually manufactured by two methods:

1. Direct incorporation method
2. Precipitation method

7.4.1   Direct Incorporation Method

According to this method the drug is directly incorporated into previously prepared vehicle. For simplification the process is divided into various steps.

Preparation of Vehicle: All the soluble components of formulation are dissolved in appropriate amount of vehicle. In case suspending agent is used, it is allowed to hydrate for sufficient time so that it is uniformly dissolved.

Preparation of Drug before incorporation: Before incorporation into vehicle, the drug powder is milled in order to get the desired particle size. Moreover, hydrophobic drug powder is thoroughly mixed with wetting agent before incorporation into the vehicle.

Incorporation of Drug into Vehicle: The drug powder or drug-wetting agent mixture is incorporated into the vehicle while continuously mixed. In case of flocculated suspension, flocculating agent is added to the suspension. After adding additional excipients, e.g., colorants or flavors, the final volume of suspension is adjusted.

Homogenization: Finally, the product is passed through appropriate mill or homogenizer in order to attain desired particle size and homogeneous dispersion. Irrespective of the manufacturing method, mixing speed should be critically adjusted. For instance, pseudoplastic (shear thinning) suspensions, e.g., flocculated suspension, typically require high-speed mixing; contrarily in dilatant (shear thickening) suspensions, high-speed mixing may increase the viscosity of suspension that will hinder efficient mixing of suspension. Therefore, in this case slow-speed mixing might be required.

Volume Adjustment: The final volume of the suspension is adjusted by adding appropriate solvent.

7.4.2 Precipitation Method

Preparation of Drug Solution: The drug is dissolved in solvent or a portion of vehicle. Co-solvents such as glycerol or ethanol may be added in case the drug has poor aqueous solubility.

Preparation of Vehicle: All the soluble components of formulation are dissolved in appropriate amount of vehicle. In case suspending agent is used, it is allowed to hydrate for sufficient time so that it is uniformly dissolved.

Precipitation of Drug: The drug solution is subjected to counterion or antisolvent that precipitates the drug into insoluble particles. Byproducts formed during precipitation might pose concerns if the concentration is too high. In such cases, the precipitated drug should be washed thoroughly and subsequently filtered.

Incorporation of Drug into Vehicle: The precipitated drug is incorporated into vehicle while continuously mixed. These suspensions are usually deflocculated and, therefore, mixed slowly.

Homogenization: Finally, the product is passed through appropriate mill or homogenizer in order to attain desired particle size and homogeneous dispersion.

Volume Adjustment: The final volume of the suspension is adjusted by adding appropriate amount of solvent.

7.5   Stability Study of Pharmaceutical Suspensions

The stability of flocculated and dispersed system is determined by following test methods which are considered very useful for determination of these parameters.

Appearance: Some characteristics are important to be observed. For instance, color and appearance at equilibrium are observed in a graduated glass cylinder or transparent glass container. Variation in color is often an indication of differences in particle size. Moreover, uniformity of sediment, presence of air pockets in the sediment, and adherence to the wall of the manufacturing vessel are also important to be observed.

Particle Size and Shape: Particle size distribution, crystal shape, and changes in physical properties can be observed by microscope. Poor distribution and differences in particle size are characterized by variation in color of suspension. To make final judgment, sufficient fields and samples need to be studied. Dilution for microscopic examination should be made with dilute dispersion medium rather than water, since change in solvent may change the crystal behavior of drug.

Recently, light-scattering techniques, such as photon correlation spectroscopy (PCS), are gaining popularity for measuring the particle size of suspension. This technique additionally gives information about the polydispersity of suspension. Other instrument options for particle size measurement are laser diffraction (LD), ultrasound attenuation (UA), and single particle optical sensing (SPOS).

Odor and Taste: Odor and taste are also important characteristics for oral suspensions. Change in odor and taste may indicate instability, change in crystal habit, and subsequent particle solubility.

Sedimentation Rate and Sedimentation Volume: Sedimentation rate and sedimentation volume are measured periodically during storage. Batch-to-batch sedimentation rate and sedimentation volume should be reproducible. Graduated glass cylinders with sufficiently wide opening are typically used to overcome the wall effect, i.e., the adhesive forces acting between the suspended particles and inner surface of cylinder.

The *degree of flocculation* (β) is the ratio of sedimentation volume (F) of the flocculated suspension to that of the deflocculated suspension. The ratio of the final volume of sediment (Vs) to the initial volume of suspension (Vi) is known as *sedimentation volume* (F). The degree of flocculation gives assessment about a point of reference, i.e., the suspension before and after flocculation.

The value of "F" may be 1, less than 1, or more than 1; for deflocculated suspension the value of F is small (i.e., less than 1), whereas for flocculated suspension, as floccules occupied the large volume, the sedimentation volume is near to 1, or in some cases more than 1, i.e., when the floccules formed are so loose and fluffy that the volume they occupy is greater than the original volume, as shown in Fig. 7.6.

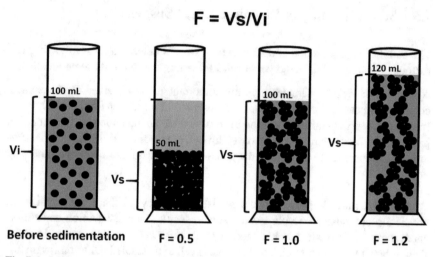

Fig. 7.6 Sedimentation volume of different types of suspension

Viscosity: The viscosity of suspension is an important parameter to be measured. Viscosity is typically measured at a certain shear rate and fixed temperature, termed as *apparent viscosity*. Brookfield viscometer is a common device used for studying the rheological behavior of suspensions.

Density: The density of suspension should be measured using a well-mixed, homogenized suspension. Precision hydrometer is a useful device for measuring the density. A decrease in density usually indicates entrapment of air in the bulk fluid.

pH Value: The suspension PH must be measured at specified temperature, after settling equilibrium is achieved. It is important to note that PH should not be adjusted using electrolytes, because it may compromise the physical stability of suspension.

Freeze-Thaw Cycling: Exposure of suspension to freeze-thaw cycling gives useful information about the physical stability of suspensions. However, for comparison a well-known resembled marketed suspension should be included in the testing, because pharmaceutical suspensions are typically not designed to withstand freezing during shelf life.

Drug Content Uniformity: Content uniformity of suspension is ascertained by measuring drug content in a "unit of use" sample (e.g., 5 mL of oral suspension) taken from top, middle, and bottom of a homogeneously mixed suspension.

Dissolution Testing: A common method for dissolution testing is submerging polyvinylidene fluoride membrane (of suitable porosity) pouch, containing known amount of suspension, in a specified dissolution medium using USP dissolution apparatus 1. Aliquots are withdrawn a predetermined interval and the amount of drug is quantified.

Other Tests: such as effectiveness of preservative, assays for potency, product compatibility with primary container-closure system, etc. should be performed in similar manner to that performed for pharmaceutical liquid solutions.

References

1. Jones DS. FASTtrack Pharmaceutics dosage form and design: Pharmaceutical Press, London; 2016. pp 25-35.
2. Nutan MTH, Reddy IK. General Principles of Suspensions. In: Kulshreshtha AK, Singh ON, Wall GM, editors. Pharmaceutical Suspensions: From Formulation Development to Manufacturing. New York, NY: Springer New York; 2010. p. 39-65.
3. Aulton ME, Taylor KMG. Aulton's Pharmaceutics E-Book: The Design and Manufacture of Medicines, Elsevier, Amstrdam. 2013:7.pp 417-425
4. Garad S, Wang J, Joshi Y, Panicucci R. Preclinical development for suspensions. Pharmaceutical Suspensions: Springer; 2010. p. 127-76.
5. Brunaugh AD, Smyth HD, Williams III RO. Essential Pharmaceutics: Springer; 2019. pp 91-110.
6. Bhattacharjee S. DLS and zeta potential–what they are and what they are not?. Journal of Control Release. 2016;235:337-351.
7. Ohshima H. Approximate analytic expression for the stability ratio of colloidal dispersions. Colloid and Polymer Science. 2014;292:2269-74.
8. Eyley S, Vandamme D, Lama S, Van den Mooter G, Muylaert K, et al. CO_2 controlled flocculation of microalgae using pH responsive cellulose nanocrystals. Nanoscale, 2015;7(34):14413-21.
9. Ferrar JA, Sellers BD, Chan C, Leung DH. Towards an improved understanding of drug excipient interactions to enable rapid optimization of nanosuspension formulations. International Journal of Pharmaceutics, 2020;578:119094.
10. Vo A, Feng X, Patel D, Mohammad A, Kozak D, Choi S, et al. Factors affecting the particle size distribution and rheology of brinzolamide ophthalmic suspensions. International Journal of Pharmaceutics. 2020;586:119495.
11. Kulshreshtha AK, Singh ON, Wall GM. Pharmaceutical suspensions: from formulation development to manufacturing: Springer; 2009. pp 20-50
12. Frigaard I. Background lectures on ideal visco-plastic fluid flows. Lectures on Visco-Plastic Fluid Mechanics: Springer; 2019. p. 1-40.
13. Hebishy E, Buffa M, Guamis B, Blasco-Moreno A, Trujillo A-J, Technologies E. Physical and oxidative stability of whey protein oil-in-water emulsions produced by conventional and ultra high-pressure homogenization: effects of pressure and protein concentration on emulsion characteristics. Innovative Food Science & Emerging Technologies,2015;32:pp.79-90
14. Loftsson T. Drug stability for pharmaceutical scientists: Academic Press, Cambridge, Massachusetts;2014. pp 50

Chapter 8
Tablets and Capsules

Amjad Khan, Majeedullah, and Saeed Ahmad Khan

Abstract Solid dosage form is estimated to be around 90% of all dosage form. Tablets and capsules are the most common solid dosage forms. The widespread use of tablets and capsule is due to their convenience of administration and ease of manufacturing. This chapter presents basics of tablets and capsules and the formulation and manufacturing technology. Moreover, the chapter also reviews the different techniques for coating of tablets. The chapter also discusses in detail the formulation and manufacturing of hard and soft gelatin capsules.

Keywords Tablet formulation · Wet granulation · Dry granulation · Tablet compression process · Direct compression · Rotary tablet press · Tablet coating · Hard gelatin capsules · Soft gelatin capsules

8.1 Tablet Dosage Form

A solid unit dosage form is prepared by molding or compression of drug(s) (active ingredient) in the presence or absence of excipients (inactive ingredients). The ingredients usually in powder form are mixed and subsequently molded or compressed into a solid dose. It may be coated to mask the bitter taste, to give a smooth texture, to improve stability, to impart special drug release characteristics, etc. Tablet is one of the most popular dosage forms. For instance, roughly two-thirds of the prescription drugs are in solid dosage forms, and half of them are in tablet form [1]. It is usually administered orally, but other routes (e.g., sublingual, rectal, or vaginal) can also be used. Ideally, tablet should be:

A. Khan · Majeedullah · S. A. Khan (✉)
Department of Pharmacy, Kohat University of Science and Technology,
Kohat, Khyber Pakhtunkhwa, Pakistan
e-mail: saeedkhan@kust.edu.pk

© The Author(s), under exclusive license to Springer Nature Switzerland AG 2022   95
S. A. Khan (ed.), *Essentials of Industrial Pharmacy*, AAPS Advances in the
Pharmaceutical Sciences Series 46, https://doi.org/10.1007/978-3-030-84977-1_8

- An elegant product having its unique identity and free from contamination, discoloration, and cracks
- Does not break during manufacturing, packaging, shipping, and dispensing
- Physically stable to maintain its physical attributes
- Chemically stable over shelf life, i.e., the ingredients do not degrade with time
- Having predictable and reproducible drug release properties

A balance between physical and chemical attributes is needed; otherwise the tablets might be having either very good physical tablets but no drug release or vice versa.

Advantages Tablets offer the highest dose precision and the least content variability than other oral dosage forms. They are economical than other oral dosage forms. Product identification is simplest and cheapest. Tablets can be easily swallowed. Special release profile can be achieved. They have ease of scalability. Tablets have better stability (mechanical, chemical, and microbiological) compared to other oral dosage form and offer flexibility in dosage.

Limitations Some drugs may not able to be compressed because of their amorphous nature or low density. Drugs with slow dissolution, poor wettability, and requiring large doses are difficult to be formulated. They are not suitable for emergency cases. Drugs unstable in GIT are not suitable for tablet formulation, e.g., protein drugs such as insulin. Additional step (coating) is required for bitter taste drugs or oxygen- or moisture-sensitive drugs, which adds up to the cost.

Some of the common types of tablets are shown in Table 8.1.

8.1.1　Excipients Used in Tablets

Compressed tablets usually contain a number of pharmaceutical adjuncts, known as excipients, in addition to the medicinal substance(s). The use of appropriate excipients is important in the development of optimum tablet formulation. Excipients are required to increase bulk of powder, to improve flow of powder, to control the rate of disintegration and dissolution, and to improve stability and compatibility of drugs. The choice of excipients is dependent on the process used for tablet manufacturing [2]. Mostly the following excipients are used in tablets. Common excipients used in tablet and their roles are summarized in Table 8.2.

Table 8.1 Types of tablets and their properties

Compressed tablets

Single compression of powder under a high pressure (~tons/in²) using compression machines

Most of the commercial tablets are prepared by single compression

Multiple compressed tablets

A portion of powder is compressed initially, followed by compression of another portion of fill material, e.g., multilayered tablet or tablet-in-tablet

Each portion of the fill material contains a different drug

Chewable tablets

Intended for disintegration in the mouth when chewed or allowed to dissolve

Commonly employed for administration of multiple vitamins, antacids, and nonsteroidal anti-inflammatory drugs

Typically contain mannitol (around 50%) as diluent. Especially useful for children

Molded tablets

Prepared by molding, i.e., forcing dampened powder into a mold of any shape

Generally reserved for laboratory- and small-scale production

Table 8.2 Excipients and their role in tablet formulation

Excipient class	Examples
Diluents or fillers	
Diluents are used (irrespective of the preparation technique) to increase tablets mass	Anhydrous lactose, microcrystalline cellulose (Avicel)
Diluents must be economical and have good compression properties. The choice of diluent depends on experience of the manufacturers' experience, its compression properties, its cost, and compatibility with other formulation ingredients	Lactose monohydrate
	Spray-dried lactose (specifically used in direct compression)
	Starch (also used as a binder and disintegrant)
	Dextrin
	Dibasic calcium phosphate
	Mannitol, etc.
Binders or adhesives	
Required to impart cohesiveness to powder for granulation and tablet integrity	Starch paste
Added as a solution or as a solid (followed by granulating fluid, usually water) into the powder	Pregelatinized starch
	Hydroxypropyl cellulose
	Hydroxypropyl methyl cellulose (hypromellose)
	Polyvinylpyrrolidone (used in conditions where water needs to be avoided)
	Polyethylene glycol 6000 (anhydrous binder, used in conditions where water and alcohol cannot be used)
	Gelatin
	Natural gums
Lubricants and glidants	
Materials that coat the surface of particles and thus:	*Hydrophobic lubricants*
Improve powder flow property (glidants) and prevent adhesion of powder to punches and dies	Magnesium stearate
Facilitate ejection of tablet from the die	Stearic acid
	Glyceryl behenate
	Glyceryl palmitostearate
	Hydrophilic lubricants
	Polyethylene glycol (PEG), e.g., PEG 4000, 6000, and 8000
	Polyoxyethylene stearates
	Sodium or magnesium salts of dodecyl sulfate
	Talc (hydrated magnesium silicate) mainly used as glidant
	Colloidal silicon dioxide, mainly used as glidant
Disintegrants	

(continued)

Table 8.2 (continued)

Excipient class	Examples
Facilitate breaking of tablet in gastrointestinal fluid	Croscarmellose sodium
Function by wetting or expanding when exposed to medium. Thus, facilitate wettability of tablet or swelling of tablet in gastrointestinal fluid	Microcrystalline cellulose
	Starch
	Microcrystalline cellulose
	Sodium starch glycolate
	Cross-linked polyvinylpyrrolidone (crospovidone)
	Pregelatinized starch
Release rate modifiers	
Polymeric materials that modify the release from tablets	Sodium alginate
	Chitosan
	Polymethyl methacrylate
Miscellaneous excipients	
Tablet formulation may also need adsorbents, sweetening agents/flavors (for chewable tablets), and colors and opacifiers	

8.1.2 Techniques for Tablet Preparation

8.1.2.1 Direct Compression Technique

Both wet granulation and dry granulation techniques require series of steps; hence both are time-consuming and relatively expensive. Another striking option for the manufacture of tablets may be direct compression, i.e., the powder is mixed and subsequently compressed into tablets, thereby avoiding the granulation step [3]. Direct compression involves the following steps:

Premilling of formulation components The size of particles in powders (drug and the excipients) is critical for compression properties of the mixture. Powder needs to be premilled (using Fitz mill, etc., discussed in Sect. 3.7.1.6) in order to ensure homogeneous mixing in subsequent steps.

Mixing of drug with excipients In this step, drug and powdered excipients are mixed. The powder mixers (previously discussed) can be be used for this process, such as planetary bowl mixer, tumbling mixers, and high-speed mixers.

Compression of powders into tablets Mixed powder is fed into the die cavity where the lower and an upper punch compresses it into tablet.

8.1.2.2 Wet Granulation Technique

Tablet manufacturing by wet granulation method has the following steps:

Dry Mixing

- The drug and excipients are mixed together. All the ingredients must be free from big lumps (usually done by prior milling).
- The mixing efficiency increases if the powders have similar particle size/distribution.
- The mixing speed and time must be validated to ensure homogeneous mixing. Generally lower shear rate is required at this stage.
- Ribbon mixers, sigma blade mixer, planetary bowl mixer, high-speed mixer-granulators, etc. can be employed for dry mixing.

Wet Mixing

- Mixed powders are mixed with a suitable binder solution or fluid (water, isopropanol, ethanol, or their mixtures) if binding agent is already included in powder mixture.
- The cohesive properties of binder enable formation of wet mass.
- Mixing continues until uniform dispersion (termed as end point) is attained. The end point of wet mixing is roughly determined by pressing a portion of wet mass in palm, and the lump of wet mass should crumble under moderate pressure. Wet massing time and moisture content determination are some of the cheap methods for determination of end point. However, in advance technology the in-built electronic/electrical devices are relying on ampere, torque, digital imaging, near infrared, acoustic sound, etc. [4].
- The mixing time needs to be validated, which depends on the wetting property of powder mixture, the granulating fluids, and mixer type.
- In modern practices, dry mixing and wet mixing steps are performed within a single operation [5]. Typically, high-speed mixers/granulators are employed for this purpose, e.g., the Diosna mixer/granulator and Collette-Gral mixer/granulator (for detail, see Sect. 2.4.6.2).
- Fluidized bed dryers are also recently used for granulation purpose. The fluidized powder mixture is simultaneously sprayed with granulation fluid containing binder.

Wet Sifting

- In wet sifting, the wet mass is passed through suitable mesh to convert it into granular aggregates.
- The purpose of wet sifting is to increase surface area to facilitate drying. However, in many instances it may be omitted.

Drying

- The wet granules are then dried to facilitate further processing.
- Drying temperature, drying time, and moisture content need to be validated.
- Different types of driers can be used for this purpose, such as tray dryer, fluidized bed dryer, etc.

Dry Sieving

- The dried granules are milled to the required particle size by passing through granulator having screens of specific pore size, e.g., oscillating granulator (discussed in Sect. 3.7.1.4). The mesh size needs to be validated (Table 8.3).

Table 8.3 Mesh size required for powder to pass thorough and the corresponding tablet sizes [6]

Tablet size	Mesh size
Up to 4.8 mm	20
Up to 7 mm	16
Up to 10 mm	14
11 mm or above	12

– The yield is determined after dry sieving.
– This step is important to control the particle size distribution, which not only improves the flow property of granules but also determines the size of tablet. In general, the granule size decreases as the tablet size decreases.

Lubrication
• Lubricants are mixed with the granules in final step before compression. It improves the flow of granules and reduces friction of tablet with punches and walls of the die.
– Same equipment can be used which were employed for mixing of powders at the start of the process. However, tumbling mixers, e.g., cone mixer and cube mixer etc. are more common (detailed in Sect. 2.4.6.2).
– Yield is calculated at this stage as well.

Tableting (Compression)
• Granules flow to the die cavity and compressed by the upper and lower punches therein.

8.1.2.3 Dry Granulation Technique

Similar to wet granulation technique, the granules are subjected to compression subsequent to preparation of granules. However, no solvent is used in dry granulation technique. High stresses are applied to powder to convert it into aggregated particles and hence granules [7]. There are two methods of formation of granules, i.e., slugging and roller compaction.

Slugging: In slugging method, the powders after mixing are compressed into a big sized tablet (termed as slugs). The slugs are milled to the required size granules using conventional milling equipment.

Roller Compaction: The drug and excipients are mixed together (similar to the previous method), subsequently pressed through two rollers (rotating in opposite direction), to produce a thin compressed sheet of material, which is then converted into desired size granules using conventional milling equipment, as described previously.

The main advantages and limitations of wet granulation and dry granulation are summarized in Table 8.4.

Table 8.4 Advantages and limitations of wet granulation and dry granulation processes

Wet granulation	Dry granulation
Advantages	*Advantages*
Reduced separation of ingredients during storage and/or processing	Common grades of excipients are used
Useful for low doses of therapeutic agent	The absence of solvent omits the use of heat
Usually, common excipients are used for manufacturing	
Tablets prepared by wet granulation are suitable for coating.	
Limitations	*Limitations*
Often involves several processing steps	Roller compaction requires special equipment for granulation
The presence of solvent may degrade the drug during processing	The components may segregate after mixing
Heat is required to remove the solvent	Tablet prepared is comparatively softer, hence having difficulty in processes after compression, e.g., coating
	There may be reduction in the yield of tablets due to over dusting

8.1.3   Tablet Compression Process

Compression originates from the Latin word *compressare*, meaning "to press together." It is an important step in tablet manufacturing, in which a pharmaceutical powder is pressed in a confined space (called die) using two metal rods (called punches). Compression usually involves two phenomena, i.e., compaction and consolidation, diagrammatically shown in Fig. 8.1.

8.1.3.1   Compaction

Compaction is the process by which the porosity of a given powder is decreased as a result of being squeezed together by mechanical means.

Pharmaceutical powder contains air spaces; therefore, the bulk volume decreases when external force is applied. The reduction in volume can be due to:

Particle close repacking: The particles rearrange due to external force; hence the bulk volume reduces.

Particles Deformation: The geometry of individual particles changes. If the deformation is reversible (i.e., particles behave like rubber), it is said to be *elastic deformation*. Though, all solids experience some extent of elastic deformation. However, in some material, the deformation is no more reversible after a certain elastic limit (yield point). This type of deformation is called *plastic deformation*, which happens in material where breaking (tensile) strength is higher than shear strength, i.e., the material squeezes rather than it breaks.

Fig. 8.1 Effect of
compression on powder

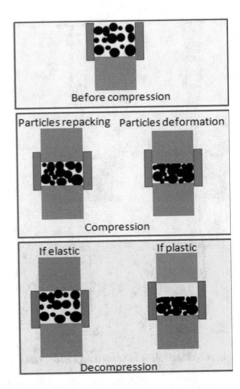

However, materials with tensile strength ⁺ shear strength (e.g., sucrose) break as a result of applied load (termed as brittle fracture). The smaller fragments fill the vacant air spaces in bulk powder.

The above mentioned phenomena contribute to the reduction in bulk volume of powder during tableting. However, in filling of hard gelatin capsules, repacking is the primary mechanism for bulk volume reduction and negligible amount of particle deformations is observed.

8.1.3.2   Consolidation

When two particles approach each other by a distance less than 50 nm, their exists a strong attractive force, and this process is called *cold welding* [8]. The bonds formed resemble those of internal structure of single particle. This is considered to be one of the mechanisms responsible for increasing mechanical strength of material during compression.

Being irregular in shape, most particles have many points of contacts. When load is applied to particles, the force is transmitted through these contact points, which results in generation of considerable amount of heat due to fraction. This consequently melts the contact area of the particles, which upon solidification results into *fusion bonding* [9]. This in turn increases the mechanical strength of material.

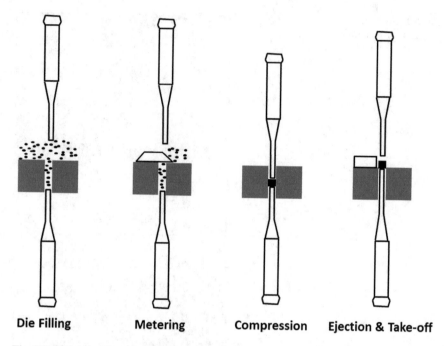

Die Filling   **Metering**   **Compression**   **Ejection & Take-off**

Fig. 8.2 Major steps in tablet compressing process

The phenomenon of cold welding and fusion bonding can be influenced by several factors, such as chemical composition of the material, available surface area, and surface contamination.

Tablet is formed by the compression of powder using two punches and die. The machine used for tablet compression is called tablet press. Tablet compressing processes (schematically shown in Fig. 8.2) can be divided into four distinct steps:

(a) Die filling
(b) Metering
(c) Compressing
(d) Ejection and takeoff

Die Filling

It involves the flow of powder from hopper into the feed frame and hence into dies. Appropriate volume of powder must be filled to ensure size, weight, and content uniformity of the tablets.

Metering

It involves the removal of excess powder so that exact weight (volume) of powder is compressed into tablets. The metering ramp (also called the dosage ramp) adjusts the position of lower punches and hence determines the amount of powder to be filled in each die. The excess powder is scraped off from the die table.

Compression
The upper and lower punches are directed toward the compression zone using cam tracks, where they are pushed by upper and lower compression rolls, hence compressing the confined powder into tablet. The thickness and hardness of tablet are determined by the distance between upper and lower punches

Ejection and Takeoff
It involves removal of tablet from the die. After compression the upper punch retracts while lower punch rises within the die cavity, hence moving the tablet upward to the surface of die table. Subsequently, the tablet is directed toward discharge chute by a scraper (also called takeoff blade).

8.1.4 Tablet Compression Machines

Pharmaceutical tablets are usually manufactured by rotary tablet presses.

8.1.4.1 Rotary Tablet Press

Tablet press can be of two types, i.e., single-punch tablet press or rotary tablet press (depending upon the number of punches it has). However, rotary presses are more common these days. The main parts of rotary tablet press are summarized in Table 8.5.

The first commercial rotary press was designed by engineer Frank J. Stokes (from Pennsylvania, USA) in the late 1800s. Stokes collaborated with Thompson and Capper (English company) to manufacture the Stokes tablet press and tooling. At the start of World War II, Stokes relocated to Pennsylvania and focused on manufacturing facilities there. The trained staff left in England started competing with Stokes in the name of Manesty Machines Ltd., which has been one of the leading manufacturers of rotary tablet presses all over the world [10].

Ever since the tablet presses have tremendously been evolved. For instance, initially, the single-punch presses could produce 100 tablets per minute, which was increased to 640 tablets per minute with rotary tablet press. Today high-speed presses are capable to produce up to 24,000 tablets per minute [11].

8.1.5 Defects in Tablets and Their Remedies

The tablets can undergo defects during compression process. The causes and remedies of problem associated with compression process are explained in Table. 8.6.

Table 8.5 Main components of rotary press

Component	Function(s)
Hopper	Containing powder
Turret	Portion of the head which holds the punches
Upper and lower Punches	Compress the powder
	Eject the compressed tablet
Feed frame	Distributes power on die table and hence into dies
Die table	Holds dies
Die	Accepts powder from feed frame and retains it until compression occurs
Scrape-off blade	Scrapes the excess powder off the die table
Feed cam	Lowers the lower punch to allow powder to come in
Compression rolls upper/lower	Force upper and lower punch to compress
Ejection cam	Lifts the lower punch and pushes tablet above the turret
Takeoff blade	Removes tablet off die table
Hydraulic	Provides hydraulic pressure
Instrumentation	Monitors force, speed, and distance
Air handling	Provides exhaust

8.1.6 Tablet Coating

Coating of pills was first introduced by Muhammad ibn Zakariya al-Razi (Rhazes) (850–923). Ibn-Sina (Avicenna) had used gold and silver coating for tablets. Gelatin coating of pills was introduced by Garot in 1838. The first film-coated tablet was marketed by Abbott in 1959, ever since a paradigm shift has been observed in coating technologies [12]. For film coating the polymer is first dissolved/dispersed in a suitable solvent and then sprayed on the tablet surface in the form of fine mist. The polymer in the form of droplets spreads over the surface of tablets and forms a thin film upon drying [13]. Tablet coating is typically performed in a rotating pan. The coating solution or dispersion is sprayed simultaneously with drying using hot air. Hot air stream removes solvent, leaving behind a thin layer of coating material on tablet surface. Therefore, some of the prerequisites are that the tablets must have sufficient hardness (not less than 8 kp) and must not chip or break during coating. The friability of tablets to be coated should not be more than 0.5%. Moreover, no formulation component should be affected by elevated temperature and humidity during the coating.

8.1.6.1 Purpose of Tablet Coating

Tablets are coated for many purposes:

- For protection of drug from extreme conditions in GIT (an enteric coating)
- For prevention of adverse effect on the stomach caused by the drugs like NSAIDs

Table 8.6 Defects in tablet compression and their remedies

Problem	Causes and remedies
Sticking and picking	*Causes*
Sticking, when powder adheres to the face of the punch tips or to the dies	Rough or damaged punches or dies
	Excessive moisture in granules
	Excessive relative humidity in environment
	Insufficient lubricant
	Insufficient compaction force
	Sharp angles in letters or logos
Picking: describes sticking within the letters, logos, etc., on punches	*Remedies*
	Polish or change punches or dies
	Dry the granules
	Dehumidify the environment
	Adjust lubricants
	Increase compaction force
	Sharp angles should be avoided in the engraving style
	"Shocking the press," i.e., the compression force is increased to make a few very hard tablets; this causes the stuck granules at the punch face to adhere with the tablet
Double impression	*Causes*
Punches having engraving on them may make new impression on the table	Free rotation of the upper or lower punch during compression
	Remedies
	Add keys on round punches
	Use punch retainers
Mottling	*Causes*
An unequal color distribution, i.e., tablets have light and dark areas	Different drug and excipient colors
	Dye may migrate to the surface of particles during drying
	Colored binder solution may not distribute evenly
	Remedies
	Use homogeneous material
	Change the solvent system
	Reduce the temperature for drying
	Overmixing or over wetting of hot paste. However, care should be taken not to affect disintegration time

(continued)

Table 8.6 (continued)

Problem	Causes and remedies
Dark spots	*Causes*
Dirt-like spots on tablets	Dirt, dust, or press lubrication
	Excessive lubrication of upper punch
	Punch tips worn
	Worn die bore
	Remedies
	Frequently clean punches
	Replace punches
	Replace dies
Chipping	*Causes*
Breaking of tablet edges	Upper punch misalignment
	Punch tips worn
	Remedies
	Replace punches
	Inspect turret for excessive punch guide wear and die pocket wear
Thickness/weight variation	*Causes*
Less/more quantity of granules is filled into the die	Too high speed of compression
	Powder lost/gained after die filling due to worn bores
	Worn scraper or improper scraper adjustment
	Nonuniform lower punch length/seat or inconsistent lower punch flight
	Dirt bellow weight adjustment
	Incorrect setting of feeder
	Inconsistent granules, i.e., variation in size distribution
	Poor powder flow (bridging or rat holing) leads to empty feed frame
	Remedies
	Adjust the speed
	Replace the dies
	Adjust/replace scrapper
	Replace lower punch/seat
	Clean the weight adjustment cam
	Adjust feeder properly
	Narrow size distribution
	Introduce vibration mechanism in the hopper Reduce the machine speed
Capping and lamination	*Causes*

(continued)

Table 8.6 (continued)

Problem	Causes and remedies
Capping, the partial or complete separation of the top or bottom of tablet from main body of the tablet	Air entrapment
	Ringed die bore
Lamination, separation of a tablet into or two distinct layers	Worn punches
	Too dry granules
	Insufficient binder
	Too much fines in the granules
	Remedies
	Reduce press speed
	Reduce fines in granules
	Reverse the dies
	Replace the worn punches
	Adjust moisture and binder ratio

- For controlling the drug release in GIT (pH-dependent coating)
- For site-specific drug targeting in GIT, e.g., colon targeted drug delivery
- For masking of bitter and nauseating taste of the drug
- For improving appearance of tablets

8.1.6.2 Types of Tablet Coating

Based on the purpose of coating and functionality of formulation components, tablet coating has been classified into sugar coating and film coating.

Sugar Coating

Sugar coating involves covering of tablets with a coat of sugar, applied as sugar solution containing other ingredients like color.

Sugar coating was introduced in the mid of 1800s [15] on the basis of the idea taken from the candy industry. Tablets were rotated in a rotating pan where a thick layer of aqueous solution of sugar was applied to them. Nowadays the sugar coating is mostly replaced by polymeric film coating. It involves several steps (Fig. 8.3):

- Sealing
- Sub-coating
- Smoothing (or grossing)
- Coloring/color coating

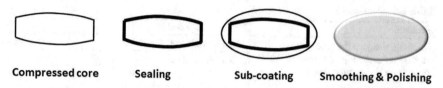

Fig. 8.3 Schematic representation of different steps involved in sugar coating process

– Polishing (or glossing)
– Printing (if needed)

Seal coating (sealing): It is performed in order to prevent moisture penetration into the tablet, since absorbed moisture might soften the tablet or disintegrate the tablet. Shellac is the most commonly applied sealant agent; however, it sometimes increases disintegration and dissolution time of tablet during storage because shellac polymerizes with time. Moreover, zein is also used as sealant.

Sub-coating: It involved the application of sticky binder solution followed by a dusting with sub-coating powder and then drying. Gelatin/acacia solution is typically used as binder. This process is repeated until the entire tablet is covered and sufficient thickness is achieved. Generally, the weight of tablet is increased by 50–100%.

Syrup (smoothing/color) coating: In this step the irregular surface of tablet is filled. Smoothing is the most technical and skillful step that involves repeated coating with syrup solution containing the dye.

Polishing: The sugar-coated tablets are subjected to lustering using beeswax or carnauba in standard coating pan or polishing pan.

Film Coating

Film coating involves application of a thin layer/coat of a polymer on to the tablets or particles. Film coating was first introduced in the 1950s and proved to be an alternative for sugar coating. It can be classified on the basis of solvent used, i.e., organic coating and aqueous coating. However, a more common classification is on the basis of its functions, i.e., nonfunctional coating (e.g., immediate-release film coating) and functional coating (e.g., modified-release film coating).

Immediate-release film coating: It is also known as nonfunctional film coating or conventional film coating and is applied only for taste masking, aesthetic purposes, product safety, and identification. It has no effect on dissolution rate of the drug.

Modified-release film coating: It is also called functional film coating. It is used for modifying the release characteristics of drug, for instance, delaying the release to surpass the deleterious effect of gastric fluid. This type of coating is called enteric coating. Similarly, extended-release film coating is intended to sustain the release

Table 8.7 Excipients used in film coating

Class of ingredients	Examples
Film former (polymers)	*For nonfunctional coating*:
	Hydroxypropyl methylcellulose (HPMC)
	Methyl hydroxyethyl cellulose
	Ethyl cellulose
	Hydroxypropyl cellulose
	Polyvinylpyrrolidone (PVP)
	Sodium carboxymethyl cellulose
	Dimethyl aminoethyl methacrylate (Eudragit E), polyethylene glycols, etc.
	For enteric coating:
	Cellulose acetate phthalate (CAP)
	Eudragit L (soluble at pH 6)
	Eudragit S (soluble at pH 7)
	HPMC phthalate, polyvinyl acetate phthalate (PVAP), etc.
	For extended-release coating:
	Eudragit® RS
	Eudragit® RL
	Eudragit® FS 30 D
Solvent	*For aqueous coating*:
	Water
	For organic coating:
	Ethanol
	Methanol
	Isopropyl alcohol
	Chloroform
	Acetone
	Methyl ethyl ketone and methylene chloride
Plasticizer	Triethyl citrate
	PEG 200
	PEG 400
	Propylene glycol
	Castor oil, glycerin, etc.
Surfactants (used for emulsion and suspension-based coating)	Polysorbates (Tweens)
	Sorbitan esters (Spans), sodium dodecyl sulfate, etc.
Coloring/opacifying agent	Titanium dioxide (most common)
	Talc
	Aluminum silicate
	Magnesium carbonate
	Calcium sulfate and aluminum hydroxide

of drug from tablet for longer time. The formulation components of film coating are given in Table 8.7.

8.1.6.3   Coating Equipment

The process of tablet coating involves rotation of tablets in a coating pan, spraying of coating solution or dispersion, and concurrent drying of tablets with hot air. Hot air stream removes solvent, leaving behind a thin layer of coating material on tablet surface. Several pans have been used.

Conventional Coating Pan

Conventional coating pan is a circular metal pan mounted at an angle that is able to rotate on its horizontal axis. The tablets are tumbled by continuous rotation of the pan, while coating solution is sprayed or ladled over the tablet bed. Hot air is supplied inside the pan through a duct and exhausted out using another duct mounted in front of the pan, as shown in Fig. 8.4.

Limitations: Conventional pans exhibit low drying efficiency, since drying is limited to the surface only. There are higher chances of dead spots due to inefficient mixing. The drying and mixing efficiency is improved by introducing additional elements into the design of the pan or air supply. For instance, *Pellegrini pan*, that has a tapered side of the pan integrated with baffles, has much improved mixing efficiency. Similarly, in another modified design, i.e., *the immersion sword system*, the drying air is introduced inside the tablet bed, using an immersed perforated metal sword-shaped device. The drying air supplied by immersed perforated sword flows upward through the tablet bed, thus providing efficient drying during coating. Likewise, modification was introduced by immersing a tube integrated with spray nozzle mounted onto the tube, i.e., *immersed tube system*. The tube supplied hot air while the nozzle sprays the coating solution onto the tablet bed simultaneously. This modification further improves the drying efficiency of conventional coating pans. Despite these modifications, coating with conventional coating pans is a slow process compared to latest coating devices.

Perforated Drum Coater

These systems utilize perforated or partially perforated coating chamber that rotates on a horizontal axis. Perforated drums provide efficient drying compared to conventional coating pans, thus requiring less time for coating. Spray coating using a perforated coating pan is illustrated in Fig. 8.5. When using this technique, the tablets are brought into a rotating drum which makes the tablets cascade down the moving

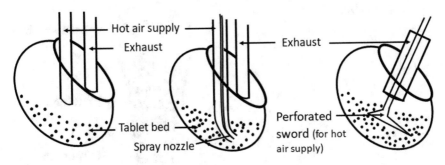

Fig. 8.4 Conventional coating pan (**a**), with immersion tube system (**b**) and immersion sword system (**c**)

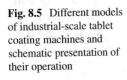

Fig. 8.5 Different models of industrial-scale tablet coating machines and schematic presentation of their operation

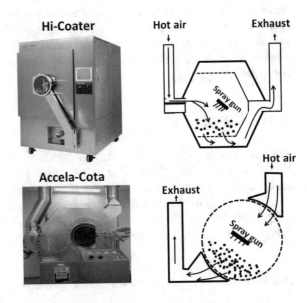

top of the bed [16]. The drum can partially or fully be perforated. Through one or multiple nozzles, the coating solution or suspension is sprayed onto the moving bed of tablets. Also an atomization air is sent through the nozzles to obtain a spray with fine droplets. Simultaneously, hot air is drawn through the tablet bed which makes the solvent of the coating solution or suspension to evaporate. These repeating cycles of coating and drying ensure a uniform coat.

Accela-Cota and Hi-Coater Systems: Both of these systems rely on the passage of air through the tablet bed continuously rotated in a perforated or partially perforated drum.

Dria coater: The air is supplied through a hollow perforated ribbed surface of the drum. Drying air passes up through the tablet bed and exhausted from the back of the pan.

Fig. 8.6 Fluidized bed
coaters

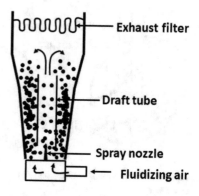

The movement of tablets in a rotating drum is less abrasive than the circulating movements they make in a fluidized bed coater, especially for large tablets since a high air velocity is required to make them circulate. However, the heat and mass transfer and the inter-tablet coating uniformity are poorer and the process takes significantly longer [17].

Fluidized Bed Coaters

They consist of a perforated plate through which air is sent into the particle bed [18]. In the center of the ground plate, a draft tube is placed, which helps in circulation of coated particles. The nozzle is placed in the center of the draft tube, so particles are sprayed, move upward in the column, and make a fountain movement to finally come back to the first point above the nozzle. This repeated movement enables to form a coherent and relatively uniform coating layer, as shown in Fig. 8.6.

8.1.6.4 Problems Associated with Tablet Coatings

Tablet coating is a technical process and all the formulation and process variables should be optimized. Any variation will lead to significant defects in final product leading to high wastage. Some of the common problems encountered during coating process are summarized in Table 8.8.

8.2 Capsule Dosage Form

Capsules contain drug that is retained in an external shell primarily made from gelatin. This gelatin shell breaks apart in the digestive tract and releases the drug that is absorbed into the blood and then circulated and processed in much the same way as APIs from a tablet [19]. In capsules a soluble gelatin container holds a dose of the

Table 8.8 Defects in tablet coating and their remedies

Coating defect	Causes/remedies
Twinning: Two tablets stick together. It is mostly observed in oblong-shaped tablets	*Causes*
	Slow evaporation of coating suspension
	Remedies
	Reduce spray rate
	Increase pan speed
Blistering: Formation of blister on the tablet surface	*Causes*
	Entrapment of gases in the film due to fast drying
	Remedies
	Use mild drying condition
Orange peel effect: The coating resembles the surface of an orange	*Causes*
	Rapid drying
	High solution viscosity
	Remedies
	Use mild drying conditions
	Use additional solvents to decrease viscosity
Filling: The monogram or bisect is filled	*Causes*
	Solution applying rate is high
	Too much coating solution
	Remedies
	Control coating solution application rate
Color variation: Visible differences in color of film found from tablet to tablet	*Causes*
	Improper mixing of color
	Nonuniform coating process
	Improper solvent chosen
	Poor spray gun setup.
	Remedies
	Proper mixing of coating solution
	Uniform spray rate
	Proper solvent selection
	Correct gun setup

drug. The drug might be in the form of a powder, a liquid or a semisolid [20]. The shell is primarily composed of gelatin and other additives such as colorants, opacifiers, plasticizers, and preservatives.

Advantages Capsules manufacturing does not need some of the unit processes that are used in tableting, such as granulation, drying, compression, etc. APIs formulated and filled as liquid in hard or soft gelatin shells can serve to increase the oral bioavailability. Soft gels are generally preferred for liquid-filled APIs. In capsules, liquids might be conveniently taken orally by patients as a unit dosage form [21]. Tampering of capsules is always difficult [22].

Limitations Capsules require specialized industrial machinery. Capsules with liquid-filled APIs have potential stability concerns on long storage. Problems concerning the uniformity of fill weight and content are also associated with capsules [21].

8.2.1 Types of Capsules

There are two types of capsules, i.e., hard gelatin capsules and soft gelatin capsules, which vary not only in their mechanical properties but also in their design.

8.2.1.1 Hard Gelatin Capsules

Hard gelatin capsules were first introduced in 1833, which were prepared using gelatin and originally consisted of two halves, a body and a cap. These components of capsules still prevail [23]. Different sizes of capsule shells are available (Table 8.9).

As already mentioned, the main constituent of hard capsule shell is gelatin. Other natural polymers, such as hypromellose and starch, have also been tested [24].

Table 8.9 Various available hard gelatin capsules, sizes, and their filling capacities

Capsule size	Capsule volume (mL)	Capacity in mg			
		Powder density			
		0.6	0.8	1.0	1.2 g/mL
000	1.37	822	1096	1370	1644
00el	1.02	612	816	1020	1224
00	0.91	546	728	910	1092
0el	0.78	468	624	780	936
0	0.68	408	544	680	816
1	0.50	300	400	500	600
2	0.37	222	296	370	444
3	0.30	180	240	300	360
4	0.21	126	168	210	252
5	0.10	78	104	130	156

"el" represent elongated sizes that are specially designed and can fill an extra 10% volume compared to standard sizes [24]

Fig. 8.7 Different types of
hard gelatin capsule shells

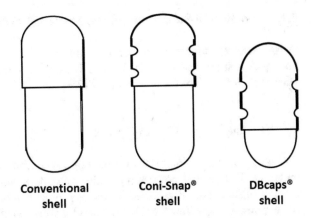

| Conventional | Coni-Snap® | DBcaps® |
| shell | shell | shell |

However, the favorable properties make gelatin as the most common material for capsule shells. Besides gelatin, the capsule shells contain ingredients like plasticizer, dyes, opacifiers, preservatives, and 12–16% water [23].

The conventional capsules are formed by the simple overlocking of caps and bodies. However, the modern shells, such as Coni-Snap® and DBcaps®, offer much improved interlocking (shown in Fig. 8.7)

Preparation of Hard Gelatin Capsule Shell

The process of capsule shell preparation is divided into the following steps [25].

Dipping: The capsule shell manufacturing machines have two sets of bars (for body and cap), each holding a series of pins that are lubricated prior to use. These cold metallic pins are dipped in the hot gelatin solution (35–40 °C). The caps and body pins are individually dipped into gelatin solutions of different colors.

Spinning: After withdrawal from the gelatin solution, these pins are rotated, to spread gelatin equally. The temperature drop of gelatin and the spinning act cause the gelatin to gel on the pin's surface uniformly.

Cooling: The hot gelatin solution is cooled down by passing through a blow of cooled air.

Drying: The coated pins are then passed through a series of a drying zone where the degree and rate of drying is carefully controlled.

Stripping: The body and caps are stripped individually from pins with bronze jaws.

Cutting: The shells are rotated and a blade is brought to bear against the cap and body. It cuts them to the appropriate lengths.

Joining: The caps and bodies are joined together to be packed and transported [26].

8.2.1.2 Industrial-Scale Filling of Hard Gelatin Capsules

During bulk manufacturing of capsules, the final formulation is packed into the body part, and the cap is pressed over it. Thus, the two halves are joined together, where the cap overlaps the body entirely [26].

Powder Filling in Hard Gelatin Capsules

Capsule formulation requires almost the same excipients that are used in tablet formulation. The general powder properties must be identified before formulating them in capsule. For instance, good flow of powder is required for efficient filling, since, powder has to flow from hopper into the capsule body. Un-uniform flow can cause variations in capsule weight. Powder clumping is also undesired during filling. Powders that are filled by dosator-type device need some degree of powder compressibility [27]. Capsule filling is a well-established technology lately. The available filling equipment ranges from small-scale manual filling, through semiautomatic filling, to large-scale fully automatic filling machines. The key steps in filling of hard gelatin capsules adapted in almost all filling machines involve the following steps [28]:

(a) Rectification of capsules (aligning empty capsules on the detachable plate so that bodies lie down driven by vacuum)
(b) Separation of caps from bodies
(c) Filling of bodies with material
(d) Rejoining of caps and bodies
(e) Ejection of filled capsules

Powder dosing is done by either dosator device or tamping device.

Dosator device: In this filling approach, the capsule caps and bodies are separated from each other and are housed in holes of their respective turntable. The turntable having bodies is rotated beneath a hopper that fills the bodies with powder [23]. The turntable containing caps is aligned over the bodies, and both capsule parts are joined together to form capsules.

The powder quantity that is dispensed to every single capsule depends on the fraction of time that the hopper spends over the capsule body (which itself is reliant on the rotary speed of the turntable). In the last part the filled capsules are pushed from the turntable with force using pins [26].

Tamping device: In this method, a plug of powder is made and is physically transferred into the capsule body. A tube with a spring-loaded piston is pushed in a powder bed. A mass of powder plugs into the tube [29]. The tube having a plug of the compacted powder is raised out of the powder bed, spun, and positioned over the capsule body. The powder plug is pushed into the capsule body by the

Table 8.10 Comparison of different powder fill hard gelatin capsule machines [12–17]

Parameters	Manual machine	Semiautomatic (dependent type)	Automatic (independent)
Fill capacity (hour)	Up to 8000	15,000–25,000	Intermittent motion (3000–60,000)
			Continuous motion (30,000–150,000)
Fill material	Powder and pellets	Powder, granules, or pellets	Powder, pellets, and granules
Capsule sizes	0 to 5	Suitable for all sizes	Suitable for all sizes
Applications	Experimental/extemporaneous filling	Industrial-scale filling	Industrial-scale filling
Examples	Jaansun	Lilly, Parke-Davis machines	Intermittent motion
			Zanasi, Pedini, Macofar, Bonapace.
			Continuous motion
			MG2, Imatic, Osaka

piston of the tube. Finally, the caps are placed on the filled capsule bodies and are ejected from the machine.

The compaction force (20–30 N) used for plug formation is much lower than the one used for compression of tablets, i.e., 10–30 kN [29]. Several models of capsule-filling machines are available that use tamping device, as shown in Table 8.10.

Liquid/Semisolid Filling in Hard Gelatin Capsules

Liquid/semisolid can also be filled in hard gelatin capsules. The liquids are either lipophilic liquids/oils enclosing dissolved/dispersed drug (e.g., sunflower oil, ara-chis oil, olive oil, and glyceryl monostearate) or water-miscible liquids containing dissolved/dispersed drug (e.g., polyethylene glycol and poloxamers) [26].

A key consideration for liquid filling in hard gelatin capsule is the choice of the solvent. When hygroscopic solvents are filled in hard gelatin capsules, it will cause splitting of the capsule shells due to excessive moisture uptake.

The liquid/semisolid formulations are filled into hard gelatin capsules by a volumetric dosing system. The filling of semisolid is typically performed at temperatures wherein the fill remains in the liquid state. Hard gelatin capsules enclosing liquid/semisolid fills are more prone to leakage, and, hence, additional treatment is required at the locking point of capsules to avoid this problem [26].

Fig. 8.8 Preparation and
filling of soft gelatin
capsules

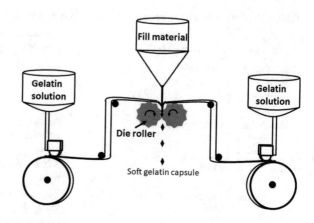

8.2.1.3 Soft Gelatin Capsules (Softgels)

Softgels are the capsules where the mechanical properties of gelatin are altered by
adding a plasticizer, thus making a very flexible sheet to be able to encapsulate liquid material. Softgels are available in a variety of shapes like round, oval, oblong,
tubular, etc. [22].

Manufacture of Soft Gelatin Capsules

Manufacturing of softgels was patented by RP Scherer in 1933. Although, the process has been modified ever since, the main principle of filling remains still the
same (Fig. 8.8). Briefly, gelatin, plasticizer(s), water, and other required ingredients
are heated together to form a viscous solution [24]. Subsequently, gelatin solution is
poured onto two rotating drums, as a result gelatin sheets are formed. Both the
sheets are fed into oppositely rotating dies that create pockets. Simultaneously, a
measured volume of fill material is added into the pocket and sealed by heat
(37–40 °C) and pressure. The capsules are torn from gelatin sheet, collected, and
washed. These capsules are dried to eliminate 60–70% w/w water before keeping
under controlled environment of humidity (20–30% RH) and temperature
(21–24 °C) [30]. After this equilibration period, the moisture inside the capsule
shell is adjusted to required limit.

References

1. Aulton, M.E. and K.M. Taylor, 2017, *Aulton's Pharmaceutics: The Design and Manufacture
   of Medicines*. 6th Edition, Elsevier, Amsterdam, Netherland: p. 501–542.
2. Liu, X., et al., *Characterization of amorphous solid dispersions*. Journal of Pharmaceutical
   Investigation, 2018. **48**(1): p. 19–41.

3. Armstrong, N.A., *Tablet manufacture by direct compression.* Encyclopedia of pharmaceutical technology, 2007. **3**: p. 3673–3683.
4. Kuriyama, A., et al., *In-line monitoring of a high-shear granulation process using the baseline shift of near infrared spectra.* AAPS PharmSciTech, 2018. **19**(2): p. 710–718.
5. Reynolds, G.K., Le, P.K., and Nilpawar A.M., *High shear granulation*, in *Handbook of powder technology.* 2007, Elsevier, Amsterdam, Netherland: p. 3–19.
6. Amin, M. and Rahman, M., *Study and impact evaluation of particle size distribution on physicochemical properties of three different tablet formulations through sieve technology.* International Journal for Pharmaceutical Research Scholars, 2014. **3**(1): p. 448–463.
7. Kittikunakorn, N., Liu, T., and Zhang F., *Twin-screw melt granulation: Current progress and challenges.* International Journal of Pharmaceutics, 2020: p. 119670.
8. Khlibsuwan, R. and Pongjanyakul, T., *Chitosan-clay matrix tablets for sustained-release drug delivery: effect of chitosan molecular weight and lubricant.* Journal of Drug Delivery Science and Technology, 2016. **35**: p. 303–313.
9. Bellini, M., Walther, M., and Bodmeier, R., *Evaluation of manufacturing process parameters causing multilayer tablets delamination.* International journal of pharmaceutics, 2019. **570**: p. 118607.
10. Natoli, D. 2008, Time for a World Tooling Standard (Accessed 28 March 2022) *https://www.pharmtech.com/view/time-world-tooling-standard.*
11. Peeters, E., et al., *Influence of extended dwell time during pre-and main compression on the properties of ibuprofen tablets.* European Journal of Pharmaceutics and Biopharmaceutics, 2018. **128**: p. 300–315.
12. Zhang, R., Hoffmann, T., & Tsotsas, E. Novel Technique for Coating of Fine Particles Using Fluidized Bed and Aerosol Atomizer. Processes, 2020. **8**(12): p. 1525. doi:https://doi.org/10.3390/pr8121525
13. Jilani, A. Abdel-wahab, M. S. and Hammad, A. H., (2017), Advance Deposition Techniques for Thin Film and Coating, In: Modern Technologies for Creating the Thin-film Systems and Coatings, IntechOpen Limited, London. DOI: 10.5772/65702
14. Parveen, S. "Recent Advancement In Tablet Coating Technology," World Journal Of Pharmacy and Pharmaceutical Sciences, 2017: p. 2189–2204.
15. Chapain C. and Stryjakiewicz, T. "Introduction – Creative Industries in Europe: Drivers of (New) Sectoral and Spatial Dynamics," Creative Industries in Europe, 2017. p. 1–15.
16. Kirollos, R. Longo, O. Brown, M. and Herdman, C. "Introducing the Head-Mounted Rotating Drum," Journal of Vision, 2016. **16**(12): p. 887.
17. Gutfinger C. and Chen, W. H. "Heat transfer with a moving boundary — Application to fluidized-bed coating," International Journal of Heat and Mass Transfer, 1969. **12**(9): p. 1097–1108.
18. A. A. Lipin and A. G. Lipin, "Prediction of coating uniformity in batch fluidized-bed coating process," Particuology, 2021. doi; https://doi.org/10.1016/j.partic.2021.03.010
19. Murachanian, D. An Introduction to Two-Piece Hard Capsules and Their Marketing Benefits. Pharmaceutical Dosage Forms. CRC Press, 2017. **30**: p. 15–30.
20. Damian, F., Harati, M., Schwartzenhauer, J., Van, CO., Wettig, SD. Challenges of Dissolution Methods Development for Soft Gelatin Capsules. Pharmaceutics. 2021. **4**;13(2): p. 214.
21. Loyd, V. A. The Art, Science, and Technology of Pharmaceutical Compounding, 6th Edition. The American Pharmacists Association. 2020.
22. Ariga, K., Ji, Q., Richards, GJ., Hill, J. P. Soft Capsules, Hard Capsules, and Hybrid Capsules. Soft Materials, 2012 **10**(4): p. 387–412.
23. Kaur, D. Formulation and evaluation of hard gelatin capsules containing Bacopa Monnieri. International Journal of Pharmaceutical Education and Research. 2020. **8**;1(02): p. 33–37.
24. Murachanian D. An Introduction to Two-Piece Hard Capsules and Their Marketing Benefits. Pharmaceutical Dosage Forms [Internet]. CRC Press; 2017. **30**: p. 15–30.
25. Lightfoot D. K. Hard Shell Capsule Filling Machines. Pharmaceutical Dosage Forms. CRC Press; 2017. **30**: p. 131–90.

26. Venkatesan, A., Radhakrishnan, A., Sahithi, K., Kuppuswamy, G. Replacement of Gelatin in capsules shell manufacturing – A Review. International Journal of Pharmaceutical Research. 2020. **1:** p. 12

27. Al-Tabakha, M. M., Arida, A. I., Fahelelbom, K. M. S., Sadek, B., Abu-Jarad R. A., et al. Influence of capsule shell composition on the performance indicators of hypromellose capsule in comparison to hard gelatin capsules. Drug Development and Industrial Pharmacy. 2015. **14(41):** p. 1726–37.

28. Prasad, V. D. Formulation and modifying drug release from Hard and Soft Gelatin Capsules for Oral drug delivery. International Journal of Research and Development in Pharmacy & Life Sciences. 2017. **6(04):** p. 2663–77.

29. Taylor, K. Aulton M. Aulton's Pharmaceutics: The Design and Manufacture of Medicines. Elsevier, Edinburgh London New York Oxford Philadelphia St Louis Sydney Toronto. 2018. p. 598–609.

30. Allen L. and Ansel H. (2014). *Ansel's Pharmaceutical Dosage Forms and Drug Delivery Systems*. Philadelphia: Lipincott Williams and Wilkins. 2014. p. 239–262

Chapter 9
Oral Inhalers

Elena Haettig and Marc Schneider

Abstract This chapter focuses on the pulmonary application of drugs, which is of great importance for the local treatment of lung-related diseases like asthma and COPD but has also been put more into the focus for systemic treatment of a multitude of diseases lately. An important aspect for inhalation is the geometry and anatomy of the respiratory tract. The system of airways, best imagined as branches, is the root of the aerodynamic properties that are used to apply drugs through inhalation and explain the various ways particles are deposited in the lungs. In order to maximize the drug delivery efficiency, the properties of the aerosol need to be optimized in order to profit off this innate aerodynamic setup. Systems to qualify aerosols related to their aerodynamic properties have been implemented but come with their downfalls, especially when it comes to transferring the results onto the human respiratory system. This needs to be taken into account when developing new inhalation devices. As the number of inhalation devices grows, it is important to understand the basic mechanism and the advantages of each system in order to optimize treatment for patients.

Keywords Aerodynamic properties · Shape factor · Aerosol generation · Pulmonal deposition · Airway geometry · Particle engineering

9.1 Introduction

Inhalation therapy dates back at least 4000 years to India where powdered Durata plants were smoked through a pipe to take advantage of their bronchodilating contents [4]. Since then, a lot of advances have been made with 1956 being described as the beginning of the modern era of inhalation therapy due to the introduction of

E. Haettig · M. Schneider (✉)
Biopharmaceutics and Pharmaceutical Technology, Saarland University,
Saarbrücken, Saarland, Germany
e-mail: Elena.Haettig@uni-saarland.de; Marc.Schneider@uni-saarland.de

the first metered dose inhaler [78]. Right now, inhalation therapy mainly focuses on local therapy for diseases concerning the lungs like asthma, COPD, or infections, and small molecules as well as macromolecules like proteins (Pulmozyme®) are used. But there is a lot of effort done in current research to use pulmonary application as a less invasive alternative to parenteral application in medication that acts systemically. So far, for systemic effect, inhalable formulations containing insulin (Afrezza® and Excurba®), loxapine as an antipsychotic (Adasuve®), or cannabis and its natural or synthetic constituents have been approved.

There are a lot of advantages of inhalation therapy in comparison to oral or parenteral application mostly due to the higher local drug concentration and reduced systemic absorption through the lung which leads to lower systemic side effects. This is of great importance for locally acting drugs like corticoids or adrenergic drugs. Since there is a fraction of the inhaled drug that will be swallowed and make its way through the intestines, using drugs with low oral bioavailability will further decrease systemic side effects. There is also lower drug metabolism [11, 16] compared to other administration routes. This is especially interesting for macromolecules like peptides and proteins which show a rapid degradation in the intestinal tract due to the high number of peptidases and proteinases. In contrast to orally administered drugs, it also bypasses the liver and with that its first pass effect. The enzymatic makeup in the lungs seems to be similar compared to that of the liver, though lower in number [10], which in return can be used as catalyzers for the activation of prodrugs [17]. Its large surface area makes it an interesting target for systemic applications. A big problem for inhalation therapy for local and systemic administration of drugs is physical barriers like clearance through mucus and macrophages or successful penetration of the surfactant and the strong dependence on patients' inhalation maneuvers.

9.2 Respiratory Tract

9.2.1 Macrostructure

The respiratory tract as a generational model was first described by Weibel and is now accepted by most scientists and can be imagined in the form of a tree starting at the trachea which forms the main airway and generation 0. In each generation the airways divide into two smaller airways. This runs from the trachea (generation 0) to the bronchi (gen. 1) and all the way to the bronchioles with the alveoli attached [87]. The alveoli are air sacs or cuplike structures where the gas exchange happens. In order for this to take place, the distance between the wall of the alveolus and the capillary, the blood-air barrier, has to be extremely close. It is because of the large number of alveoli that the surface area of the lungs is so extensive, which increases the amount of gas that can diffuse between the capillaries and the lung. It is similar to the effect of the villi in the intestines. The alveoli start to form at the 17th generation, and with every generation more and more alveoli emerge. The total number of alveoli differs from source to source, and a lot of textbooks claim a number of 200–300 million [30, 40], while newer calculations put the number at around 500 million, depending on the method used to count the alveoli [68]. The generational model, and with that the airways, is completed at generation 23 with the last alveoli.

The characteristics of each of the generations also have an impact on how we administer pharmaceuticals via inhalation. Velocity of the air is an important aspect to consider in understanding how the particles are carried through the airways influencing deposition. The velocity increases through the first four generations due to decreasing overall diameter in accordance with Bernoulli's principle. Bernoulli's principle suggests that velocity of a fluid in a tube is inversely proportional to its diameter. This is also valid for air as flowing gases are also considered fluids. The maximum velocity is reached in the third generation [30], where the area of the total cross section is the lowest [87]. Afterward the cross-sectional area increases with each generation and in return the air velocity decreases. In the alveolar duct and sacs, the air is nearly not moving anymore [27], changing the dominant deposition mechanism for particles.

The airways can be sectioned into a conducting and a respiratory region. The conducting region is made up of generation 0–16 and is not participating in the gas exchange. The main function of this region is connecting the external environment to the respiratory zone. While doing so, it also conditions the air to optimize gas exchange by increasing the humidity and temperature of the incoming air [45]. Due to the higher velocity and its bifurcations, it also filters bigger particles and ejects them with the help of mucus that is mostly found in this region of the lung. The respiratory region begins with the first appearance of alveoli. It consists of respiratory bronchioles, alveolar ducts, and alveoli. This is the region where the gas exchange happens.

The exact area of the lung surface is subject to big discussion. The estimates vary widely depending on the method used for calculation. Weibel, known for his description of the lung branches as generations, estimated the lung surface to be 130 m^2 with the help of stereological histology and electron microscopy [88], while other estimates say that 1 m^2/kg of lung surface seems to be found in most mammals [52], which matches the estimates of other working groups. Another problem to consider is that the surface area of the lungs changes throughout the phases of ventilation. During inhalation the surface area increases while it decreases during exhalation. But either way, the surface area that can work as an absorption area is probably in a similar order of magnitude or even larger than that of the intestines, which has been recently estimated to be around 35 m^2 [41], explaining the interest of research in using the lungs as a possible target area for systemic drug delivery. Summarizing this, the overall area of the lung varies depending on the publication from 70 to 140 m^2. In contrast, the conducting airways only sum up to an area of ~2 m^2 [36].

9.2.2   Microstructure

The main purpose of the epithelium is to separate the external from the internal environment of our body. This important task protects the systemic bloodstream from substances that are toxic to our organs. This is possible due to the tight junctions found in between epithelial cells. But next to protecting the body from toxic substances, the main function of the lung epithelium is to remove carbon dioxide from the bloodstream and saturating it with oxygen which is then carried everywhere in the body and makes the function of all organs possible.

The epithelium in the airways consists of a variety of different types of cells. The composition of epithelium is characteristic for the specific region in the respiratory tract. Ciliated cells are found in the conducting region that move mucus upward through coordinated movement of their cilia, which can be imagined as hairlike structures on the luminal surface of the cells [90]. Other cells in the epithelium of the conducting region include club and goblet cells. Goblet cells secret the mucus [74] and therefore provide an important defense mechanism of the lungs.

The two cell types compromising the epithelium of alveoli are alveolar type I cells that make up around 97% of cells in the alveolar epithelium [39] and alveolar type II cells. Type I cells are found on the luminal surface of the alveoli and are responsible for the gas exchange between the respiratory tract and the bloodstream. The diffusion of O_2 and CO_2 is possible due to a partial pressure gradient across the tissue. O_2 goes from 100 mmHg in the alveoli to 40 mmHg in the pulmonary capillaries, while CO_2 goes from 40 mmHg to 45 mmHg the other way round [30]. To provide little resistance to the diffusion of oxygen and carbon dioxide, the type I cells are extremely thin (~100 nm) facilitating and guaranteeing the main function of the lung: to supply the body with oxygen.

Alveolar type II cells secret surfactant. Surfactant is a lipoprotein complex made up of different phospholipids, neutral lipids, and proteins [21] that reduce the surface tension on the alveoli walls which lowers the work of breathing and prevents the collapse of alveoli at the end of expiration. Surfactant also serves as an additional barrier to the systemic circulation and is therefore an additional obstacle when administering drugs that are supposed to act systemically. Surfactant may lead to aggregation of proteins [17] and peptides which would hinder the systemic uptake and make them more prone to endocytosis by macrophages. Alveolar type II cells are also responsible for the regeneration of the epithelium by serving as progenitor cells and differentiating into type I cells if those have been damaged [6, 26].

Smooth muscle cells line part of the airways from trachea to the terminal bronchioles [1] and are separated from the epithelium by a connective tissue called the lamina propria. Their task in healthy individuals is not fully understood yet, but they are thought to be responsible for the bronchoconstriction in chronic obstructive diseases like asthma and COPD [2, 49]. And they are usually targeted by bronchodilating agents given in inhalation therapy.

As mentioned above, mucus is secreted from the goblet cells and mainly consists of mucins and proteoglycans [55]. It is tasked with the hydration of the epithelium to make sure it does not dry out. On top of that it also humidifies the air to close to 100% humidity [45] until it reaches the alveoli. Mucus also serves as a defense mechanism in two ways. It contains lysozyme and other defense proteins and peptides as an active way to protect the body from infections [47, 77]. Passively, the mucociliary clearance is responsible to remove foreign materials which are trapped in mucus and transported toward the trachea through the coordinated movements of cilia of ciliated cells and get either swallowed or ejected. Mucus also plays a part in different diseases, most noteworthy being cystic fibrosis wherein the composition of mucus is pathologically viscous leading to infections and an inability of the lungs to function properly [32].

Next to mucociliary clearance there are other defense mechanisms within the respiratory tract. One being the macrophages in the alveoli. There is no mucociliary clearance in the alveoli because mucus would hinder the gas exchange, so alveolar macrophages are the primary way of defense for the body. As in every other part of the body, they phagocytose pathogens and deposited particles, release antimicrobial substances, and control inflammation processes through the release of cytokines. Thus they also play a role in diseases like asthma and COPD and may play a minor part in the remodeling processes seen as sequelae of these diseases [46, 83].

9.3 Mechanism of Particle Deposition

Generally, the inhaled particles are carried by the inhaled stream of air through the airways. Due to different forces acting upon the particles and especially the inert mass, their direction can differ from the air path. To get an understanding of how particles behave, the forces present in the respiratory system need to be considered.

The first force is the drag force; it acts opposite to the relative motion of the particle [27]. As with every other fluid, the Stokes law also applies to the stream of inhaled air, which leads to the following definition of the drag force F_D:

$$F_D = 3\pi\eta d * \left(\vec{v}_p - \vec{v}_f\right)$$

(η, viscosity of air; d, particle diameter; $\vec{v}_p$, velocity of particles; $\vec{v}_f$, velocity of fluid/air)

The Stokes law only applies if the relative velocity of the air on the particle surface is zero. This is not the case if the observed particles are smaller than 10 μm [18]. In that case the drag force is divided by the Cunningham slip correction factor (C_s) which leads to the following equation: F_D/C_s

The other two forces acting upon a particle are the gravitational force F_G which is determined by the particle's mass

$$F_G = m_p * g$$

and a stochastic force F_t [81], which represents the collisions leading to the Brownian motion. These are collisions of the surrounding fluid with the observed particle. The collisions and therefore the Brownian motion are temperature dependent and correlate positively, meaning it increases with increasing temperature.

The sum of these three forces is the motion of a particle in the airstream that travels through the lung and is brought together in the Langevin equation [13]:

$$m_p * \frac{dv_p}{dt} = -F_D + F_G + F_t$$

Depending on the part of the respiratory tract and the size of the particles, the importance of the respective forces shifts, and this leads to different ways the particles are deposited.

The main deposition mechanisms for inhalation therapy are impaction, sedimentation, and diffusion.

9.3.1 Impaction

Based on Newton's laws of motion, an object is going to stay in its current movement unless a force from outside is going to act on it. This is called inertia and means that the particles are going to stay on their existing trajectories rather than following the airstream. If the deviation from the airstream gets too strong, the particles will collide with the airway wall [22] get deposited. This way of deposition is called *impaction* [27]. This is described by the first term, the drag force, of the Langevin equation. This often occurs at or near bifurcations due to the fast changes in the air streamline direction and mostly happens in the first few generations as a result of the higher velocity and the higher possibility of turbulences compared to the following generations [69, 81]. Impaction is dependent on a particle's Stokes number (Stk). The Stk suggests that two particles with different properties might still have the same airborne behavior:

$$Stk = \frac{\rho d^2 v}{18\eta L}.$$

(ρ, particle density; d, particle diameter; v, speed of the air; η, viscosity of air; L, characteristics of passage, e.g., diameter)

Rather than looking at the actual diameter, the Stk number indicates that a combination of size and density is more conclusive about the airborne motion of a particle than size or mass. This relationship can be expressed by the aerodynamic diameter ($d_æ$), which is derived from the Stk number [33]:

$$d_{ae} = \left(\frac{\rho}{\rho_{Ref}} \right)^{0,5} * d.$$

(ρ, particle density; ρ_{ref}, density of a reference material (commonly 1000 kg/m³); d, particle diameter)

If the Stk number is small, the particles are expected to follow the airstream more closely because the necessary force to overcome the inertia is smaller. The larger the aerodynamic diameter, the more likely it is for particles to get deposited by impaction. This mostly concerns particles with an aerodynamic diameter larger than 5 µm.

9.3.2 Sedimentation

Sedimentation is the most effective deposition mechanism for inhaled pharmaceuticals in the respiratory region [81]. Since most particles larger than 5 µm have been deposited by impaction in the upper parts of the airways, this mode of deposition mostly concerns particles in the range of 1–8 µm [27]. The range of particle sizes differs throughout publications, which illustrates that the exact deposition mechanism is influenced by a variety of circumstances regarding the patient (size, gender, anatomy, the breathing patterns), and the properties of the particles (size distribution, shape, surface structure, and more) and size ranges will only give estimates of the most likely way a particle is deposited. Further toward the end of the respiratory tract, drag force still comes into play, but the gravitational force gains importance due to the decreasing air velocity in the smaller airways. With lower velocities, the momentum and Stk number decrease and with that does the chance of impaction. Rather, the particles deposit from nearly not moving air due to gravitational forces. Since gravitational deposition is a time-dependent process, its importance also decreases with higher breathing rates. It is obvious from this consideration that the breathing maneuver, the strength of inhalation, breath holding, and strength of exhalation will impact on the deposition.

9.3.3 Diffusion

As mentioned above, Brownian motion is described as the motion of particles or molecules due to collisions with surrounding particles and molecules. Next to the temperature dependency, an aspect to consider is the mobility of the particles within the surrounding. This mobility is determined by the viscosity and the radii of the particles. This was described by multiple scientists [31, 80, 84] independently from each other and is now known as the Einstein relation. This leads to a diffusion coefficient representing this phenomenon:

$$D = \frac{C_S * k_B * T}{6\pi * \eta * r}$$

(D, diffusion coefficient; k_B = Boltzmann's constant; T, temperature; η, dynamic viscosity of the fluid; r, radius of the particle; C_s, Cunningham slip factor) [81]

As seen already for the drag force, the Cunningham slip factor has to be added into the equation. *Diffusion* in the context of particle deposition means that particles move through Brownian motion in the air and make random contact with the airway wall. It is a purely statistical deposition mechanism. Diffusion is important for very small particles of 1 μm and smaller. For particles this small, the gravitational force can be neglected [81]. It only happens in the alveoli when the air practically does not move, because only then the motion initiated by the collisions between the particles is not superimposed by convective motion. As with sedimentation the number of particles deposited increases with decreasing breathing rates [44] because the air is stationary for a longer period, providing the particles with more time to come in contact with the alveolar walls.

On top of that there are other mechanisms like interception which concerns particles with an elongated shape like asbestos fibers; in that case the particles do not leave the airstream, but due to their size they still get close enough to interact with the airway wall and get deposited. Furthermore, electrostatic interactions that only concern particles with a surface charge can also play a role. And these mechanisms can become important points to consider dependent on the properties of the administered particles.

9.4 Particle Engineering

When engineering particles for inhalation therapy, there are different factors to contemplate when it comes to particle properties.

Size is the largest factor to a successful delivery to the lungs. As discussed in the previous chapter, size often determines which part of the respiratory tract is reached by a particle. But it has to be recognized that different drugs that act locally might not have the same ideal deposition area in the respiratory tract. Rather, the ideal deposition area is dictated by the drug target. If targeting specific areas that have a higher density of the specific receptors would be pursued, it could lead to a far more efficient pharmaceutical therapy. Unfortunately, even for historically long targeted receptors like b-adrenoreceptors, there has not been much research into the regional distribution within the lungs [17]. Generally, 1–5 μm is regarded as best for delivery in the lungs. Here the aerodynamic diameter is always in focus and not the geometric size of the particles as described above. With decreasing size of particles, there is higher surface energy and that comes with an increased risk of agglomeration changing pulmonary deposition.

When talking about stability of size, another factor to consider is hygroscopicity of the microparticle material. In DPI there typically is a prolonged storage in powder form which could lead to an uptake of water and in return to higher agglomeration. On top of that, the humidity increases within the respiratory system which may increase the size and change the density of the particles [44, 81] leading to different flight properties and as a result to a less ideal deposition mechanism and area. While the storage stability mostly concerns DPI, the growth of particles during their travel through the respiratory system is of importance for pMDI as well.

But the aerodynamic diameter is not only comprised of size, as shown above density is also a factor. A great example from nature is pollen. These rather large particles are actually able to travel quite deep into the lungs. This is explained by their low density [18, 27]. A particle with twice the diameter but ¼ of density has the same aerodynamic size and would behave the same way in a fluid, in this case air, according to Stokes. Due to the larger size and increased surface area, the tendency of the particles to agglomerate will decrease [27], solving a recurring problem of particle engineering for inhalation therapy.

While talking about particles for inhalation therapy, the focus is usually on spherical particles. There are lots of advantages that can explain that circumstance; spheres do not have a lot of contact area which reduces the tendency to aggregate and yield good flow properties. They are also rather easy to produce in industry-scale quantities through milling or spray drying. But there are other shapes that have properties of interest for pulmonary application. Elongated, fiber-like particles align according to their shape with the airflow. Thus, the relevant parameter for the aerodynamic behavior is the diameter rather than the length of the particles. Furthermore, fiber-like particles have shape factors >1 reducing the aerodynamic diameter in comparison to spheric particles. In consequence, they often travel deeper into the lung. A good example of that behavior are asbestos fibers. The toxic mechanism behind asbestos is attributed to their shape in combination with their bio-resistance and the consequent retention time of those fibers in the lung. There is research done to use these properties in order to optimize the release profiles in pulmonary application [61].

9.5   Analysis of Aerodynamic Particle Properties

As discussed before, the aerodynamic diameter is an important factor in predicting how a particle is going to be deposited in the airway. In order to assess the aerodynamic particle size distribution (APSD), a multitude of apparatuses and techniques have been developed and tested. The US and EU Pharmacopeias primarily focus on an impact-based measurement by a so-called cascade impactor [58].

A cascade impactor (illustrated in Fig. 9.1) can be imagined as a tower of multiple stages build upon each other representing generations, though it has less than 24 stages. The inhalation device is fixed onto a tube at the top. At the bottom a vacuum pump draws the air in a consistent velocity through the different stages.

Fig. 9.1 Sketch of the principal setup of a cascade impactor. The blue line indicates the airstream initiated by the vacuum pump on the bottom of the system. The air flows through the connecting tubes from plate to plate with increasing velocity

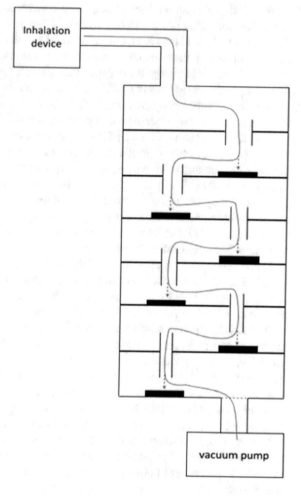

Each stage is connected to the following with a straight tube that gets narrower by each stage. This increases the velocity of the air traveling through the tubes. The placement of this tube changes on each stage. Underneath the exit of each tube is an impaction plate for the particles that are not able to follow the airstream, thus impacting on the plate. As the air carrying the particles streams through the different stages with increasing velocity due to the decreasing diameter of the tubes, it gets harder for the particles to follow the air path. This holds true especially for larger masses. Additionally, a pre-separator can be mounted before the first stage in order to remove big particles [73], like lactose carrier particles so the results in the following stages are not distorted. Each stage has a specific cutoff which is defined by the diameter of spherical particles with a specific density (commonly 1 kg/L) that will deposit with a 50% probability. This is also depicted in Fig. 9.2.

From the deposition pattern on the different stages, the resulting mass distribution is connected to the size of the particles by the mass median aerodynamic

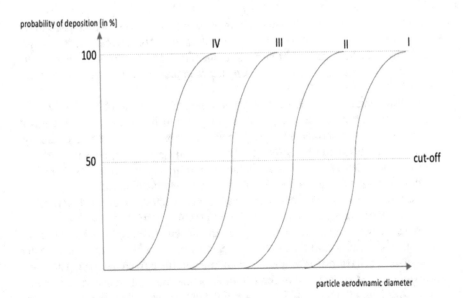

Fig. 9.2 Deposition probability of particles with a certain aerodynamic diameter in cascade stage I–IV

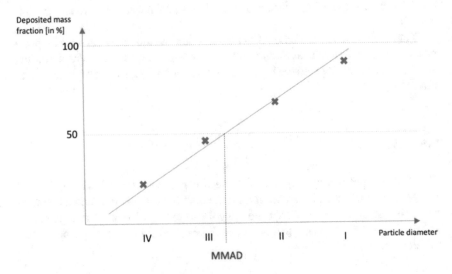

Fig. 9.3 Graphic evaluation of MMAD with the cutoff of stages I–IV of a cascade impactor

diameter (MMAD) splitting the deposited particles in two parts of equal weight. A graphical evaluation is illustrated in Fig. 9.3.

The next-generation impactor [56] (NGI) generally works the same way. But rather than having air travel vertically downward, it travels in a zigzag motion up and down horizontally facilitating handling such as sample collection and cleaning.

The impact plates might be small pools of liquid rather than solid plates. This is mostly used to determine the concentration of drug that gets deposited in each stage. The ability to further analyze the concentration of active pharmaceutical ingredients is a major advantage that cascade impactors/impingers have over other techniques [59].

Generally, particles for inhalation are not monodisperse, so rather than getting one aerodynamic diameter, the particles are going to be distributed over several stages leading to an aerodynamic particle size distribution. As mentioned already, this size distribution can be described by a common parameter, the mass median aerodynamic diameter. The MMAD is the aerodynamic diameter at which 50% of the mass is made up of particles with a lower aerodynamic diameter.

The tested inhalation device determines which impactor is to be used, how many stages, and whether a pre-separator is required by the pharmacopeias.

Though appreciated for its simplicity as a quality control apparatus, its simplicity might also be the downfall for cascade impactors as a research tool to determine the aerodynamic behavior of inhalation formulations in the clinical context. The main points of critique are the vast difference of structure to the anatomy of human airways, especially interindividual variety and the consistent airflow which is not represented in the inhalation process of humans. This holds especially true for the glottis/throat which shows huge variety and also is vastly changing during childhood. Therefore, the Alberta Idealized Throat was established trying to address this part [76].

Other techniques to measure APSD that are not mentioned or favored by pharmacopeias are measurement through the time-of-flight method or laser diffraction. The latter is especially suitable for aqueous preparations as is the case in nebulization [58].

9.6 Devices

The market of pulmonary drug devices is steadily growing, from $19.6B in 2010 [7] to $38.1B in 2017 [8]. This makes it an interesting field for pharmaceutical companies to invest into further research and bring forth new innovations. There are different devices to inhale pharmaceuticals. Each designed for different purposes and different target patients.

9.6.1 Pressurized Metered Dose Inhalers

The pressurized metered dose inhaler (pMDI) was first developed in 1956 (Medihaler®; Riker Laboratories, Inc) [37] and changed the world of aerosol inhalation. It still has a big role in inhalation therapy though innovations are now mostly found in other areas.

9.6.1.1 Device Structure

pMDI consists of a canister that is, in most cases, made out of aluminum [65] and often coated internally to reduce interactions between formulation and the canister wall [19], a metering valve, and an actuator. For protection and better handling, it is usually placed in a plastic case.

The metering valve is the most important part of the pMDI. It is responsible to keep the doses of the released aerosol consistent. To ensure this, a metering chamber with a specific volume is part of the metering valve. This chamber stores the next dose until needed. Once the patient activates the release, the metering valve opens toward the exterior and releases the dose and afterward refills the chamber again to be ready for the next dose. This is possible because once the dose is released and the valve stem closes the path to the outside again, a pressure gradient forms in which there is close to atmospheric pressure within the metering chamber and significantly higher pressure in the canister [19]. Therefore, the formulation is drawn into the metering chamber and stays there until the next release is activated.

Before the propellants were changed from chlorofluorocarbons (CFCs) to hydrofluorocarbons (HFCs) (reasons below), patients were able to check how empty their pMDI was by putting it in a bowl of water. If it was floating on the surface, it was empty. This floating test or counting the already administered doses and subtracting it from the total available doses were the only ways to determine remaining doses. After the introduction of HFCs, the float test was not possible anymore which led to complaints by MDI users and ultimately resulted in the introduction of dose counters [50]. These dose counters might show a specific number or indicate a color pattern whether a new inhaler should be acquired.

As with every other instance of drug packaging, all components of the device have to be inert to the formulation and be able to withstand the high pressure.

9.6.1.2 Formulation

The pMDI formulation consists of a propellant, an active drug, and excipients like co-solvents and surfactants [19]. The propellants are vapors with a boiling point below room temperature. Due to the pressure inside of the canister, there is an equilibrium of the vapor and its liquid phase. Consequently, a consistent pressure in the canister is ensured, which in return guarantees a consistent aerosol generation and a consistent repeated released dose (first dose is equal to last dose ejected). If the gas could not be liquified by increasing pressure but one would simply compress the gas, an aerosol would still be formed at activation, but with each dose that is administered, the pressure in the canister would decrease [62]. And due to the dose optimally aerosolized being dependent on the pressure currently present in the canister, the deposited doses would decrease over time. In the case of an equilibrium of vapor and liquid phase being present, once a dose is administered, part of the liquid phase evaporates and the pressure is automatically adjusted to the previous pressure (Fig. 9.4).

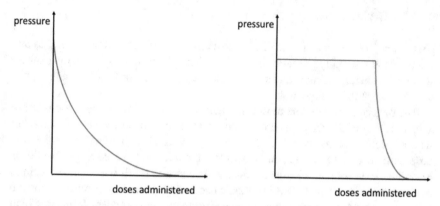

Fig. 9.4 Pressure in canister filled with compressed gas (left) and liquified compressed gas (right)

The active ingredient is dissolved or suspended in the propellant. When the release is activated, the high pressure created by pushing the fluid through the nozzle leads to the formation of small droplets. Due to the low boiling point of the propellant, the propellant evaporates once it leaves the canister which reduces the size of the droplet.

The main propellants in inhalers used to be chlorofluorocarbons (CFCs), but due to their ozone-depleting nature, their use has been banned in the Montreal Protocol of 1989. That brought forth the use of hydrofluorocarbons (HFC), also called hydro-fluoroalkanes (HFA). Today HFC-134a (1,1,1,2-tetrafluoroethane) and HFC-227 (1,1,1,2,3,3,3-heptafluoropropane) are the most commonly used propellants for pMDIs. In comparison to CFCs, HFCs do not contain chlorine and do not add to the depletion of ozone in the atmosphere. But they are still very potent greenhouse gases. This causes the future of HFCs in inhalation therapy to be questionable.

A big disadvantage of HFCs compared to CFCs is that a lot of drugs are not soluble in the propellant, so co-solvents are necessary making the formulation more complex [19]. With both CFCs and HFCs, a so-called Freon effect (named after the trade name of several halocarbons) is observed, in which the patient stops inhaling once the propellant reaches the mouth and airway because of a cold sensation on the mucosa [9].

9.6.1.3 Actuation

Pressurized metered dose inhalers can be divided into coordination and breath-actuated devices. For the coordination devices the patient is required to breathe in at the exact time as they activate the release of the aerosol to ensure a sufficient number of particles reach the lungs. A lot of patients, especially those of very young and old age, have difficulties coordinating the actuation, possibly leading to insufficient drug application. The breath-actuated devices like the Autohaler® were developed to solve these difficulties [9, 67]. These devices get activated once the patient breathes in and a vacuum is formed at the mouthpiece, a click sound signaling the release of a dose.

9.6.1.4 Spacers

In a lot of cases, the use of a spacer is recommended to overcome or at least reduce those problems with pMDIs. A spacer is a wide tube that is mounted onto the MDI device. A lot of spacer devices also offer the option to add a mask on the mouthpiece of the spacer. If a spacer is used, the particles are inhaled from standing air rather than a fast stream of air. That means that there is no coordination required; the dose is ejected into the spacer and can be inhaled through multiple breaths. In addition, large particles are already deposited in the spacer which reduces the undesired deposition in oropharyngeal region [28, 29] and consequently possible side effects in the upper airways. Because of this, the use of a spacer is always indicated when glucocorticoids are administered. Once a problem with spacers that has since been addressed by newer devices was electrostatic precipitation on the spacer walls [5], but due to modern devices made with materials without electrostatic potential, this problem could be reduced [12, 79]. For young children and in some cases also adults and elderly patients, a face mask increases the inhalation efficiency if they are unable to produce a tight seal around the mouthpiece [43, 60].

9.6.2 Dry Powder Inhalers

Dry powder inhalers (DPI) originally gained traction as an alternative for the pMDIs after CFCs were banned in 1989 [23]. Previous DPIs were unattractive due to poor drug delivery efficiency [25]. Only when alternatives were desperately needed, advances in design would make DPIs a viable option next to MDIs and nebulizers. Besides circumventing the use of CFCs or HFCs, an additional advantage of DPIs is that there is no coordination from the patient needed between actuation and inhalation. All inhalers are breath-actuated, decreasing the chance of improper inhalation due to coordination difficulties.

The success of DPIs is based on three aspects: the design of the device, the formulation, and the patient's inhalation effort [35].

9.6.2.1 Design of DPI Devices

In comparison to pMDIs, there is a vast variety of different devices with different dosing mechanisms with new ones being patented and marketed constantly. The general categorization is into single-dose and multidose devices. Single-dose devices are fed capsules filled with one dose each. In order to release the powder, the capsules get perforated, usually by piercing the capsule through pressing buttons

Aggomerates (Soft pellets)   adhesive mixture (interactive mixture)

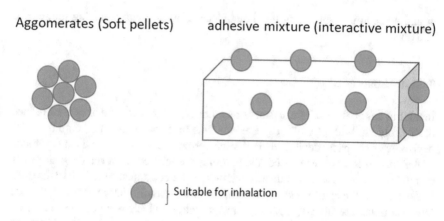

                    Suitable for inhalation

Fig. 9.5 Sketch of a soft pellet from particles suited for inhalation (left) and of an adhesive mixture having the micronized drug adsorbed on a carrier system (often lactose crystals)

on the outside of the device that are connected to small needles on the inside. Multidose inhalers, on the other hand, have a reservoir of multiple doses already present in the inhaler, either as a blister (e.g., in Diskus®, GSK) or a powder reservoir (e.g., in the Turbuhaler®, AstraZeneca). In inhalers with a powder reservoir, the continued flowability of the powder has to be guaranteed, especially with respect to humidity.

The particles in the powder are either present as agglomerates or adhesive mixtures. Both systems are designed to release the individual particles until they reach the airways (Fig. 9.5)

Therefore, the biggest problem to be addressed when designing an inhaler is the sufficient dispersion of particles by the airstream. Inhalers are designed to increase the dispersion force in different ways depending on the individual model. There are different strategies like a circulation chamber that makes sure that only smaller particles can leave the nozzle as seen in the NEXThaler® or a narrow spiral-shaped channel like the Turbuhaler® has [19, 89], in which the bigger particles get deposited on the inner wall of the channel due to impaction and many more. In single-dose inhalers, the dispersion of the particles is facilitated by the movement of the pierced capsule during inhalation.

The design of each device reflects on the airflow resistance and varies widely between the different devices [48]. Responsible for this is the need for local pressure drop or high air velocities that ensure that the full dose is carried by the air and sufficient deagglomeration has taken place [35]. De Boer [24] was able to show that a higher airflow resistance did not actually relate to higher work of breathing as one would expect. Which airflow resistance is most comfortable for patients is disputed [3, 14, 24].

Regardless of dispersion technology, a big influence on the drug delivery efficiency with DPIs that cannot be influenced by the design is the inhalation effort by the patient. This is a big disadvantage of DPIs over MDIs (though it is important to note that a minimum inspiratory flow rate is required during pMDI use as well, estimated at 20 L/min [38]). The dependency on a patient's ability to inhale becomes a problem once their breathing is temporarily or permanently obstructed, so either in an acute asthma attack or patients with severe asthma/COPD. Contrary to what one would expect, devices with higher airflow resistance require lower inspiratory flow rates to reach ideal drug release [15]. The reason for this lies in the mechanism of powder deagglomeration. Higher-resistance devices rely on the airflow resistance within the device to deagglomerate the drug powder. In low-resistance devices, the lack of resistance-induced turbulences leads to the inhalation airflow rate being the main cause of redispersion. Subsequently, a higher inspiratory flow rate is needed for devices with lower intrinsic resistance [20]. Pulmicort Turbuhaler® being the exception due to having a high resistance and a high inspiratory flow rate [54]. The lowest required flow rate which often depends on the patient's lung condition and which can be of great importance for the success of inhalation therapy, varies greatly in between devices. There is no data bank filled with the exact flow rate for each device, but Haidl et al. put together a list with already available data of previous publications. Often the values differ between publications. The results are shown in Fig. 9.6 [38]. If possible, the flow rate necessary to be achieved by the patient for a specific device was distinguished into insufficient, acceptable but room for improvement and optimum flow rate.

The variety of devices does not only present in their physical mechanism to redisperse agglomerates or their air flow resistance, but it also leads to very different modes of operation. This poses a problem if devices are switched and can lead to mishandling unless an adequate training for the patient has taken place [70].

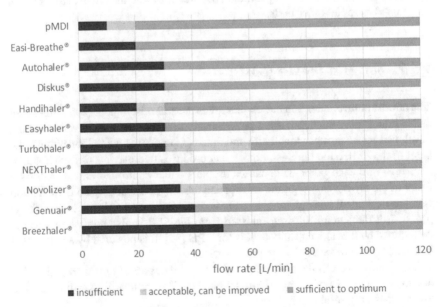

Fig. 9.6 Lowest required inspiratory flow rate with different devices to achieve insufficient, acceptable, or sufficient drug delivery to the lungs. (Adapted from Haidl et al. [38] (permission obtained))

9.6.2.2 Formulation

In order for the drug to reach the desired lung region, the drug particles should be in the range of 1–3 μm. This can be achieved by micronizing larger particles or growing them to the desired size by spray drying, for example. Due to their small size, inhalation drug particles tend to agglomerate due to their high surface energy. As a result, deagglomeration is an important process to consider, and designing the device, as investigated above, accordingly has a big influence on its success. The other option of ensuring small particle size even after longer storage is mixing coarse carrier particles with the small drug particles which are prone to aggregation (so-called interactive mixture). In most cases, α-lactose is used as the carrier particles to which the smaller particles adhere to, mostly due to Van der Waals forces. During the inhalation process, the coarse particles separate and the small drug particles are released without any further chance to aggregate. The lactose, depositing in the oropharynx, can irritate the throat leading to coughing. At the same time, the sweet taste often indicates to the patient that the dose has been released. With the absence of the Freon effect, this is reassuring to many patients.

If a biodegradable polymer were to be used as carriers of drug particles, controlled release in the airways could be possible.

A way to bypass the problem of aggregation is the "storage" of the particles as a monolithic tablet. Before use, a dose will be created by scraping off a controlled amount of drug. This system was first introduced by the MAGhaler®, also known as Jethaler® [25]. Due to the short time between the preparation of the particles and the application, aggregation is less prevalent, but breathing into the inhaler still needs to be avoided as for all other DPIs.

9.6.3 Soft Mist Inhaler (SMI)

The newest innovation in inhalation therapy is the soft mist inhaler (sometimes also called liquid spray inhalers). It is a propellant-free, multidose reusable inhaler. First developed by Boehringer Ingelheim and marketed under the name "Respimat", the heart of the SMI is the uniblock nozzle that produces two fine jets of drug solution angled toward each other. This causes droplets under controlled conditions. As a result of the lower velocity of the aerosol and the longer dispensing time, inhalation is not as dependent on simultaneous actuation and inhalation as is the case with the pMDI. But the amount of drug released from the SMI is also consistent and not determined by the lung capacity as it is for DPIs [86]. It is powered by a spring that is wound up mechanically by the patient before use.

The disadvantage of the SMI is the small range of possible fluids that can be used due to the proneness of the nozzle to get blocked. This limits the possible fluids to low viscous drug solutions but not dispersions.

Next to Respimat®, two other systems, the AERx system and Medspray nozzle, are in clinical trials. Both are based on the principle of Rayleigh breakup for the generation of droplets [51].

9.6.4 Nebulizers

As with SMIs, nebulizers form small droplets of the drug solution or suspension in order to reach the deeper lung areas, but in contrast to SMI, a nebulizer dose is usually administered in multiple breaths. The big advantage of nebulizers compared to other inhalation devices is that even high doses of drugs can be aerosolized [69] and can also be used for off-label use of other drug solutions.

There are different types of nebulizers that are characterized by their mechanism to form these small droplets.

9.6.4.1 Jet Nebulizers

The first and oldest one is the jet nebulizer which uses a compressor in order to compress gas or air that streams through a narrow pipe (Fig. 9.7). At the orifice of the pipe, as a result of the high velocity and the sudden change of gas pressure leading to high shear stress, the gas breaks up the fluid brought there by a feeding tube into a polydisperse mix of aerosolized droplets. Due to the high velocity, the bigger droplets collide with a baffle that is placed some distance from the orifice and fall back down in the drug solution reservoir [71].

Fig. 9.7 Schematic of a jet nebulizer

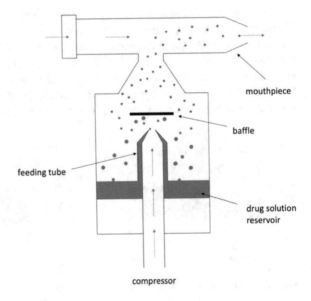

mouthpiece

baffle

feeding tube

drug solution reservoir

compressor

This can pose a problem because part of the solvent will evaporate while the droplets are suspended in the air, leading to an increasing concentration of the drug solution [57]. The smaller particles, as discussed before, are more likely to follow the air stream around the baffle and exit the nebulizer through the mouthpiece. Jet nebulizers have a multitude of disadvantages like the need for a compressor, the orifice that is suspectable to wear off due to the pressure of the compressed air or the cleaning procedures [64]. Additionally, the high shear stress caused by the air velocity might lead to degradation of the drug, especially for stress-susceptible drugs like proteins [42] or nucleic acids [53].

9.6.4.2 Ultrasonic Nebulizers

The preferred nebulizers use ultrasonic vibrations caused by a piezoelectric transducer to aerosolize the liquid which removes the need for compressed air/gas. The first of its kind, appropriately called the "ultrasonic nebulizer," uses the vibrations to form droplets on the surface of a drug solution reservoir. The exact mechanism of droplet generation is disputed [71, 91]. This leads to polydisperse droplets as well, the larger ones recover back to the reservoir due to gravity and baffles installed within the device, while the smaller ones get carried by the stream of air through the mouthpiece into the airways of the patient. The piezoelectric transducer is either in direct contact with the liquid containing the drug or is separated from it by a liquid interface. The liquid interface prevents the drug solution from overheating [34].

9.6.4.3 Mesh Nebulizers

The second kind of nebulizer using a piezoelectric transducer is the mesh nebulizer. The drug solution or suspension does not have an open surface but rather is directly covered by a mesh. There are two different setups used in these mesh nebulizers. One being the passive way (Fig. 9.8), in which a thin layer of the drug solution or suspension is placed in between the piezoelectric transducer and a stationary mesh. The vibrations of the transducer are conducted through the fluid containing the drug leading to fluctuating surface levels at the mesh. Every time the fluid presses against

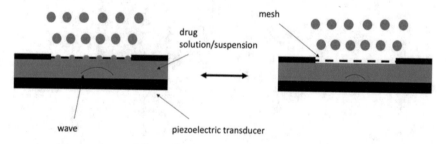

Fig. 9.8 Passive setup; at different phases of oscillation

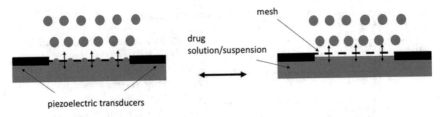

Fig. 9.9 Active setup; at different phases of oscillation

the mesh, small droplets exit through the holes [71]. According to Rayleigh's theory, the droplets released are approximately twice the size of the holes in the mesh [72].

The active setup (Fig. 9.9) has the transducer mounted onto the mesh, which means the vibrations are not carried by the fluid but by the mesh itself. The vibrations lead to a periodically changing height of the mesh relative to the fluid surface and the mesh basically pushing the fluid through its holes releasing the droplets [66].

In both cases, the droplet size is directly connected to the hole sizes of the mesh. If the holes in the mesh are comparable in size, it can be expected that the droplets are as well.

There are several situations where nebulizers present as the best option for inhalation therapy, especially if high drug doses have to be delivered or for very young and old patients since there is no specific inhalation technique required. They are also often used by patients with cystic fibrosis or pulmonary hypertension [71]. But nebulizers have the big disadvantages to not be as portable as their counterparts even after the introduction of portable nebulizers like PARI eFlow® and having a strict cleaning regimen that often puts them in the shadow of the other inhalers.

9.7 Choice of Inhaler

In order to choose the best inhaler for a specific patient, the health and ability of the patient as well as the properties of each inhaler device has to be taken into account. The pros and cons of the different inhalation device systems [9, 75, 82] are described in Table 9.1.

There are a plethora of flow charts and questionnaires developed to simplify the decision of which inhaler is the best choice for a specific patient [63, 82, 85]. While they all differ slightly, most of the times the same factors are examined: availability of drug, age of the patient, state of consciousness, ability to inhale spontaneously, coordination skills, and possible inhalation flow rate. An example of such a flowchart is represented in Fig. 9.10.

Table 9.1 Advantages and limitations of inhalation device systems

Device	Advantages	Limitations
pMDI	Portable	Coordination required
	Multidose	Propellant required
	Short treatment time	Large amount of deposition in oropharyngeal region
	Inexpensive	--Same mode of action in most devices
	Less dependent on inhalation effort of patient	
	Consistent dosing	
DPI	Portable	Dependent on inhalation maneuver
	Short treatment time	May not be possible in emergency
	Breath-actuated	Risk of agglomeration
	Dose indicator	Devices cannot be exchanged without training
	No propellant	Cannot be used by young children
Nebulizers	No specific technique required	Not as portable
	For all ages	Some need outside energy source
	Unconscious patients	Long treatment times
	High doses can be aerosolized	Extensive cleaning regimens
		Expensive
SMI	Portable	Only few drugs available
	Multidose	Not breath-actuated
	Less dependent on inhalation effort of patient	
	Less coordination required	
	Dose indicator	
	No propellant	
	No spacer needed	

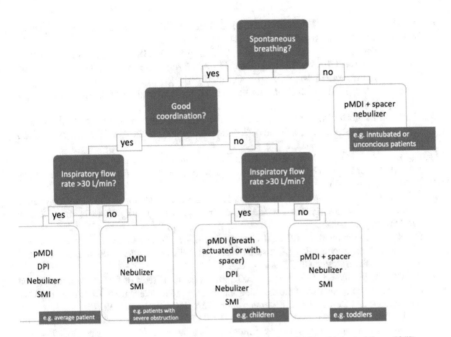

Fig. 9.10 One possible flowchart to determine ideal inhalation device. (Adapted from [85])

References

1. Amrani, Y. & Panettieri, R. A., 2003. Airway smooth muscle: contraction and beyond. *Int J Biochem Cell Biol,* 35(3), p. 272–276.
2. An, S. S. et al., 2007. Airway smooth muscle dynamics: a common pathway of airway obstruction in asthma. *Eur Respir J,* 29(5), pp. 834-860.
3. Andersen, P. B. & Hanssen, N. C. G., 1993. Which magnitude of inhaler resistance is preferred by patients using dry powder inhalers. *Eur Resp J,* p. 148.
4. Anderson, P., 2005. History of aerosol therapy: liquid nebulization to MDIs to DPIs. *Respir Care,* 50(9), 1139–1149.
5. Anhøj, J., Bisgaard, H. & Lipworth, B. J., 1999. Effect of electrostatic charge in plastic spacers on the lung delivery of HFA-salbutamol in children. *Br J Clin Pharmacol,* 47(3), pp. 333-336.
6. Barkauskas, C. E. et al., 2013. Type 2 alveolar cells are stem cells in adult lung. *J Clin Invest,* 123(7), p. 3025–3036.
7. BCC Publishing, 2012. *Pulmonary Drug Delivery Systems: Technologies and Global Markets (accessed March 10th 2021).* [Online] Available at: https://www.bccresearch.com/market-research/healthcare/pulmonary-drug-delivery-systems-hlc094a.html
8. BCC Publishing, 2018. *Pulmonary Drug Delivery Systems: Technologies and Global Markets (accessed March 10th 2021).* [Online] Available at: https://www.bccresearch.com/market-research/healthcare/pulmonary-drug-delivery-systems-technologies-and-global-markets.html
9. Bell, J. & Newman, S., 2009. The rejuvenated pressurised metered dose inhaler. *Expert Opin Drug Deliv,* 4(3), pp. 215-234.
10. Benford, D. J. & Bridges, J. W., 1986. Xenobiotic metabolism in lung. *Progress in drug metabolism,* 7(8), pp. 53-94.

11. Brown, D. T., Marriot, C. & Beeson, M., 1983. Antibiotic binding to purified mucus glycoproteins. *J Pharm Pharmacol,* Volume 35, p. 80.

12. Burudpakdee, C., Kushnarev, V., Coppolo, D. & Suggett, J., 2017. A Retrospective Study of the Effectiveness of the AeroChamber Plus® Flow-Vu® Antistatic Valved Holding Chamber for Asthma Control. *Pulm Ther.,* 3(2), pp. 283-296.

13. Chandrasekhar, S., 1943. Stochastic problems in physics and astronomy. *Rev Modern Physics,* 15(1), p. 1–91.

14. Clark, A. R. & Hollingworth, A. M., 1993. The relationship between powder inhaler resistance and peak inspiratory conditions in healthy volunteers--implications for in vitro testing. *J Aerosol Med,* 6(2), pp. 99-110.

15. Clark, A. R., Weers, J. G. & Dhand, R., 2020. The Confusing World of Dry Powder Inhalers: It Is All About Inspiratory Pressures, Not Inspiratory Flow Rates. *J Aerosol Med Pulm Drug Deliv,* 33(1), pp. 1-11.

16. Crooks, P. A. & Damani, L., 1989. Chapter 3: Drug application to the respiratory tract: Metabolic and pharmokinetic considerations. In: *Respiratory Drug Delivery.* Boca Raton: CRC Press, pp. 61-90.

17. Crooks, P. A., Penthala, N. R. & Al-Ghananeem, A. M., 2019. Drug Targeting to the Lung: Chemical and Biochemical Considerations. In: *Pharmaceutical Inhalation Aeorosol Technology.* Boca Raton: CRC Press, pp. 29-78.

18. Crowder, T. M., Rosati, J. A. & et al., 2002. Fundamental Effects of Particle Morphology on Lung Delivery: Predictions of Stokes' Law and the Particular Relevance to Dry Powder Inhaler Formulation and Development. *Pharmaceutical Research,* 19(3), pp. 239-245.

19. Da Rocha, S. R., Bharatwaj, B. & et al., 2019. Pressurized Metered-Dose Inhalers. In: *Pharmaceutical Inhalation Aerosol Technology.* Boca Raton, USA: CRC Press, pp. 427-453.

20. Dal Negro, R. W., 2015. Dry powder inhalers and the right things to remember: a concept review. *Multidiscip Respir Med,* 10(1), p. 13.

21. Daniels, C. B. & Orgeig, S., 2003. Pulmonary surfactant: the key to the evolution of air breathing. *News Physiol Sci,* Volume 18, p. 151–157.

22. Darquenne, C., 2012. Aerosol Deposition in Health and Disease. *J Aerosol Med Pulm Drug Deliv,* 25(3), p. 140–147.

23. De Boer, A. H. & Grasmeijer, F., 2019. Dry Powder Inhalation. In: *Pharmaceutical Inhalation Aerosol Technology.* Boca Raton, USA: CRC Press, pp. 455-472.

24. De Boer, A. H., Winter, H. M. I. & Lerk, C. F., 1996. Inhalation characteristics and their effects on in vitro drug delivery from dry powder inhalers Part 1. Inhalation characteristics, work of breathing and volunteers' preference in dependence of the inhaler resistance. *Int. J. Pharm,* 130(2), pp. 231-244.

25. De Boer, A. H. et al., 2017. Dry powder inhalation: past, present and future. *Expert Opin. Drug Deliv,* 14(4), pp. 499-512.

26. Desai, T., Brownfield, D. & Krasnow, M., 2014. Alveolar progenitor and stem cells in lung development, renewal and cancer. *Nature,* 507(7491), p. 190–194.

27. Desai, P. P., Mapara, S. S. & Patravale, V. B., 2018. Crystal Engineering: Upcoming Paradigm for Efficacious Pulmonary Drug Delivery. *Current Pharmaceutical Design,* 24(21), pp. 2438-2455.

28. Dissanayake, S. & Suggett, J., 2018. A review of the in vitro and in vivo valved holding chamber (VHC) literature with a focus on the AeroChamber Plus Flow-Vu Anti-static VHC. *Ther Adv Respir Dis,* Volume 12.

29. Dolovich, M., Ruffin, R., Corr, D. & Newhouse, M. T., 1983. Clinical evaluation of a simple demand inhalation MDI aerosol delivery device. *Chest,* 84(1), pp. 36-41.

30. Ehmke, H., 2016. Atmung. In: *Duale Reihe Physiologie.* Stuttgart: Thieme.

31. Einstein, A., 1905. Über die von der molekularkinetischen Theorie der Wärme geforderte Bewegung von in ruhenden Flüssigkeiten suspendierten Teilchen. *Annalen der Physik,* 322(8), pp. 549-560.

32. Fahy, J. V. & Dickey, B. F., 2010. Airway Mucus Function and Dysfunction. *N Engl J Med,* 363(23), pp. 2233 - 2247.

33. Finlay, W. H., 2019. Aerosol Physics and Lung Deposition Modeling. In: *Pharmaceutical Inhalation Aerosol Technology.* Boca Raton, USA: CRC Press, pp. 81-91.

34. Flament, M. P., Leterme, P. & Gayot, A., 2001. Study of the Technological Parameters of Ultrasonic Nebulization. *Drug Dev. Ind. Pharm,* 27(7), pp. 643-649.

35. Frijlink, H. W. & De Boer, A. H., 2004. Dry powder inhalers for pulmonary drug delivery. *Expert Opin Drug Deliv,* 1(1), pp. 67-86.

36. Fröhlich, E., Mercuri, A., Wu, S. & Salar-Behzadi, S., 2016. Measurements of Deposition, Lung Surface Area and Lung Fluid for Simulation of Inhaled Compounds. *Front Pharmacol.,* Volume 7, p. 181.

37. Grossman, J., 1994. The evolution of inhaler technology. *J Asthma,* 31(1), pp. 55-64.

38. Haidl, P. et al., 2016. Inhalation device requirements for patients' inhalation maneuvers. *Respir Med,* Volume 118, pp. 65-75.

39. Haies, D. M. & Weibel, E. R., 1981. Morphometric study of rat lung cells. I. Numerical and dimensional characteristics of parenchymal cell population. *Am Rev Respir Dis,* 123(5), p. 533–541.

40. Hedenstierna, G. & Borges, J. B., 2016. Normal physiology of the respiratory system. In: *Oxford Textbook of Critical Care.* Oxford, UK: Oxford University Press.

41. Helander, H. F. & Fandriks, L., 2014. Surface area of the digestive tract - revisited. *Scand. J. Gastroenterol.,* 49(6), p. 681–689.

42. Hertel, S. P., Winter, G. & Friess, W., 2015. Protein stability in pulmonary drug delivery via nebulization. *Adv Drug Deliv Rev,* Volume 93, pp. 79-94.

43. Hess, D. R., 2008. Aerosol Delivery Devices in the Treatment of Asthma. *Respir Care,* 53(6), pp. 699-725.

44. Heyder, J., 2004. Deposition of inhaled particles in the human respiratory tract and consequences for regional targeting in respiratory drug delivery. *Proc Am Thorac Soc,* 1(4), p. 315–320.

45. Hickey, A. J. & Thompson, D. C., 2019. Physiology of the Airways. In: *Pharmaceutical Inhalation Aerosol Technology.* Boca Raton, USA: CRC Press, pp. 5-27.

46. Hough, K. P., et al. 2020. Airway Remodeling in Asthma. *Frontiers in medicine,* Volume 7, p. 191.

47. Knowles, M. R. & Boucher, R. C., 2002. Mucus clearance as a primary innate defense mechanism for mammalian airways. *J Clin Invest,* 109(5), p. 571–577.

48. Krueger, P., Ehrlein, B., Zier, M. & Greguletz, R., 2014. Inspiratory flow resistance of marketed dry powder inhalers (DPI). *Eur. Respir. J.,* Volume 44, p. 4635.

49. Lambert, R. K. et al., 1993. Functional significance of increased airway smooth muscle in asthma and COPD. *J Appl Physiol,* 74(6), p. 2771–2781.

50. Lavorini, F., Fontana, G. A. & Usmani, O. S., 2014. New Inhaler Devices - The Good, the Bad and the Ugly. *Respiration,* 88(1), pp. 3-15.

51. Leiner, S. et al., 2019. Soft Mist Inhalers. In: *Pharmaceutical Inhalation Aerosol Technology.* Boca Raton, USA: CRC Press, pp. 493-507.

52. L'Enfant, C., 2000. Discovery of endogenous surfactant and overview of its metabolism and actions. In: *Lung Surfactants, Basic Science and Clinical Applications.* New York City, USA: Marcel Dekker, Inc., pp. 119-150.

53. Lentz, Y. K., Worden, L. R., Anchordoquy, T. J. & Lengsfeld, C. S., 2005. Effect of jet nebulization on DNA: identifying the dominant degradation mechanism and mitigation methods. *J Aerosol Sci,* 36(8), pp. 973-990.

54. Levy, M. L. et al., 2019. Understanding Dry Powder Inhalers: Key Technical and Patient Preference Attributes. *Adv Ther,* 36(10), pp. 2547-2557.

55. Lillehoj, E. P. & Kim, K. C., 2002. Airway mucus: its components and function. *Arch Pharm Res,* 25(6), p. 770.

56. Marple, V. A., Roberts, F. J., Romay, F. J. & et al., 2003. Next generation pharmaceutical impactor (a new impactor for pharmaceutical inhaler testing). Part I: Design. *J Aerosol Med,* 16(3), pp. 283-299.

57. Mercer, T. T., Tillery, M. I. & Chow, H. Y., 1968. Operating characteristics of some compressed-air nebulizers. *Am Ind Hyg Assoc J,* 29(1), pp. 66-78.

58. Mitchell, J., 2019. Aerodynamic Particle Size Testing. In: *Pharmaceutical Aerosol Inhalation Technology.* Boca Raton: CRC Press, pp. 541-587.

59. Mitchell, J. P. & Nagel, M. W., 2003. Cascade impactors for the size determination of aerosols from medical inhalers: Their uses and limitations. *J Aerosol Med,* 16(4), pp. 341-377.

60. Mitchell, J. P. & Nagel, M. W., 2007. Valved holding chambers (VHCs) for use with pressurised metered-dose inhalers (pMDIs): a review of causes of inconsistent medication delivery. *Prim Care Respir J,* 16(4), pp. 207-214.

61. Möhwald, M. et al., 2017. Aspherical, Nanostructured Microparticles for Targeted Gene Delivery to Alveolar Macrophages. *Adv Healthc Mater,* 6(20).

62. Myrdal, P. B., Sheth, P. & Stein, S. W., 2014. Advances in metered dose inhaler technology: formulation development. *AAPS PharmSciTech,* 15(2), pp. 434-455.

63. National Institute for Health and Care Excellence, 2021. [Online] Available at: https://www.nice.org.uk/guidance/ng80/resources/inhalers-for-asthma-patient-decision-aid-pdf-6727144573

64. Nerbrink, O. & Dahlbaeck, M., 1994. Basic nebulizer function. *J Aerosol Med,* 7(Suppl 1), pp. 7-11.

65. Newman, S. P., 2005. Principles of metered-dose inhaler design. *Respir Care,* 50(9), pp. 1177-1190.

66. Newman, S. & Gee-Turner, A., 2005. The omron microair vibrating mesh technology nebuliser, a 21st century approach to inhalation therapy. *J Appl Ther Res,* 5(4), pp. 29-33.

67. Newman, S. P., Weisz, A. W., Talaee, N. & Clarke, S. W., 1991. Improvement of drug delivery with a breath actuated pressurised aerosol for patients with poor inhaler technique. *Thorax,* 46(10), pp. 712-716.

68. Ochs, M. et al., 2004. The Number of Alveoli in the Human Lung. *Am J Respir Crit Care Med.,* 169(1), p. 120–124.

69. Pirozynski, M. & Sosnowski, T. R., 2016. Inhalation devices: from basic science to practical use, innovative vs generic products. *Expert Opin. Drug Deliv,* 13(11), pp. 1559-1571.

70. Price, D., 2005. The way forward: dry powder inhalers should only be switched with physician agreement and patient training. *Int. J. Clin. Pract,* Issue 149, p. 36.

71. Pritchard, J. N., Von Hollen, D. & Hatley, R. H. M., 2019. Nebulizers. In: *Pharmaceutical Aerosol Inhalation Technology.* Boca Raton, USA: CRC Press, pp. 473-492.

72. Rayleigh, J. W., 1878. On the stability of jets. *Proc London Math Soc,* s1-10(1), pp. 4-13.

73. Roberts, D. L. & Marple, V. A., 2000. USA, Patent No. 6595368B2.

74. Rogers, D., 1994. Airway goblet cells: Responsive and adaptable front-line defenders. *Eur Respir J,* 7(9), p. 1690–1706.

75. Rogliani, P. et al., 2017. Optimizing drug delivery in COPD: The role of inhaler devices. *Respir Med,* Volume 124, pp. 6-14.

76. Ruzycki, C. A., Martin, A. R. & Finlay, W. H., 2019. An Exploration of Factors Affecting In Vitro Deposition of Pharmaceutical Aerosols in the Alberta Idealized Throat. *J Aerosol Med Pulm Drug Deliv,* 32(6), p. 405–417.

77. Schutte, B. C. & Paul B. McCray, Jr. 2002. β-Defensins in Lung Host Defense. *Annual Review of Physiology,* Volume 64, pp. 709-748.

78. Stein, S. W. & Thiel, C. G., 2017. The History of Therapeutic Aerosols: A Chronological Review. *J Aerosol Med Pulm Drug Deliv,* 30(1), p. 20–41.

79. Suggett, J. et al., 2015. Use of valved holding chambers without pre-conditioning and the influence of anti-static materials. *J Aerosol Med Pulm Drug Deliv,* pp. 4-5.

80. Sutherland, W., 1905. LXXV. A dynamical theory of diffusion for non-electrolytes and the molecular mass of albumin. *Philosophical Magazine,* 9(54), pp. 781-785.

81. Tsuda, A., Henry, F. S. & Butler, J. P., 2013. Particle transport and deposition: basic physics of particle kinetics. *Compr Physiol,* 3(4), p. 1437–1471.

82. Usmani, O. S., 2019. Choosing the right inhaler for your asthma or COPD patient. *Ther Clin Risk Manag,* Volume 15, pp. 461-472.

83. Van der Veen, A., De Groot, L. E. & Melgert, B. N., 2020. The different faces of the macrophage in asthma. *Curr Opin Pulm Med,* 26(1), pp. 62-68.

84. Von Smoluchowski, M., 1906. Zur kinetischen Theorie der Brownschen Molekularbewegung und der Suspensionen. *Annalen der Physik,* 326(14), p. 756–780.

85. Voshaar, T., App, E. M., Berdel, D. & et al., 2001. Recommendations for the choice of inhalatory systems for drug prescription. *Pneumologie,* 55(12), p. 579 – 586.

86. Wachtel, H., Kattenbeck, S. & Dunne, S., 2017. The Respimat® Development Story: Patient-Centered Innovation. *Pulm Ther,* Volume 3, pp. 19-30.

87. Weibel, E. R., 1963. *Morphometry of the Human Lung.* Berlin: Springer Verlag.

88. Weibel, E. R., 2009. What makes a good lung?. *Swiss Med Wkly,* 139(27-28), p. 375–386.

89. Wetterlin, K., 1979. USA, Patent No. 4137914.

90. Yaghi, A. & Dolovich, M. B., 2016. Airway Epithelial Cell Cilia and Obstructive Lung Disease. *Cells,* 5(4), p. 40.

91. Yeo, L. Y., Friend, J. R., McIntosh, M. P. & al., e., 2010. Ultrasonic nebulization platforms for pulmonary drug delivery. *Expert Opin Drug Deliv,* 7(6), pp. 663-679.

Chapter 10
Topically Applied Products

Sajid Khan Sadozai, Arsh Zafar, and Sheheryar Sajjad

Abstract This chapter reviews the basic concepts involved in topically applied pharmaceuticals for local delivery as well as transdermal delivery. The principle of transdermal drug delivery system is to deliver drug through the skin in a controlled and predetermined rate for systemic action while dermal drug delivery is intended for local action. Stratum corneum, the upper layer of the skin plays a pivotal role as rate-limiting barrier in drug permeation. The chapter discusses how the drug permeates through the skin. Semisolid preparations for dermal drug delivery i.e., ointments and creams are discussed in detail since these are common conventional dosage forms administered topically. Moreover, transdermal patches and microneedles are also explained in this chapter.

Keywords Dermal and transdermal drug delivery · Skin · Ointments · Creams · Pastes · Gels · Permeation · Transdermal patches · Transdermal microneedles

10.1 Introduction

The three basic functions of the skin are to provide protection, regulation, and sensation. It provides a promising route for drug delivery [1]. Drug administered through the skin provides numerous advantages, such as it can be easily administered without the aid of anyone, improves patient compliance and drug delivery to targeted site, prolongs release of drug, achieves systemic (transdermal) and local (dermal) action, reduces potential systemic side effects, and eliminates GI tract problems and hepatic first-pass metabolism [2–4].

S. K. Sadozai (✉) · A. Zafar · S. Sajjad
Department of Pharmacy, Kohat University of Science and Technology, Kohat, Pakistan
e-mail: sajidsadozai@kust.edu.pk

© The Author(s), under exclusive license to Springer Nature Switzerland AG 2022
S. A. Khan (ed.), *Essentials of Industrial Pharmacy*, AAPS Advances in the
Pharmaceutical Sciences Series 46, https://doi.org/10.1007/978-3-030-84977-1_10

Drug delivery through the skin is categorized into two main groups. Drug products which are designed to achieve local action are termed as topical preparations or dermal drug delivery system whereas those products in which systemic effect is desired are termed as transdermal medications or transdermal drug delivery system (TDDS) [5]. Transdermal medications are designed to increase the amount of drug that can cross the skin barrier using various permeation enhancers and modern technology potentiate the drug to enter the systemic circulation and show systemic effects rather local effects where it is applied [6].

To better understand this difference it is important to understand the structure of the skin and its barrier function.

10.2 Structure of the Human Skin

The skin is largest organ of the body. It receives around one third of blood circulation. The skin serves as first line of defense, i.e., to provide a permeability barrier that prevents the absorption of certain biological and chemical agents [7]. The skin is composed of several layers made up of different types of cells (Fig. 10.1).

Epidermis is the outermost layer of skin. It consists of several types of cells, such as keratinocytes, melanocytes (for melanin production), Langerhans cells (antigen-presenting dendritic cells), and Merkel cells (sensory mechanoreceptors). Epidermis is divided into various layers (strata) that vary in structure, function, and barrier properties. The different layers of epidermis are stratum corneum, stratum lucidum, stratum granulosum, stratum spinosum, and stratum basale. All these layers work together to continuously rebuild skin surface [8].

Dermis lies beneath the epidermis and connected to epidermis by channeled junction called dermo-epidermal junction. Dermis is composed of many types of cells, such as fibroblasts (synthesize extracellular matrix) and mast cells (important component of innate immune system). Moreover, sweat glands, sebaceous glands, hair follicles, lymph vessels, blood vessels, and cutaneous sensory nerves are also embedded in the dermis [9].

Hypodermis is also called subcutaneous tissue, lies beneath dermis and supports the dermis and epidermis. Hypodermis stores fats, regulates temperature, and provides nutritional support and mechanical protection.

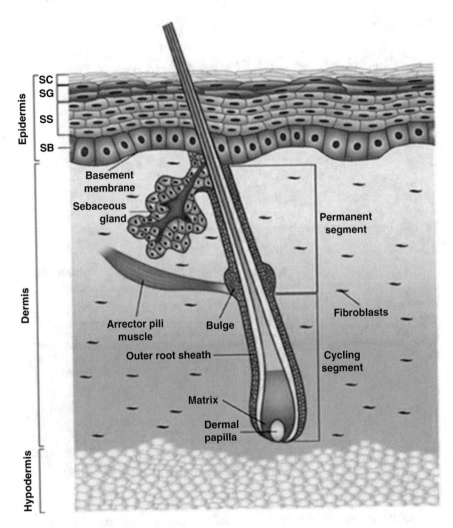

Fig. 10.1 Anatomy of the skin [10]

10.3 Dermal and Transdermal Drug Delivery

When drug delivery through the skin is intended for localized effect, it is called *dermal drug delivery* to avoid the systemic absorption of drug. Some may be intended to interact with stratum corneum e.g., moisturizers. Others may be intended to remain on skin surface e.g., sunscreens or barrier creams. Some products should reach the dermis or epidermis layers e.g., antimicrobial agents, local anesthetics, or photodynamic therapy agents [4].

For *transdermal drug delivery* drug passes through various layers of the skin to reach systemic circulation. TDDS is fabricated in such a way that it penetrates

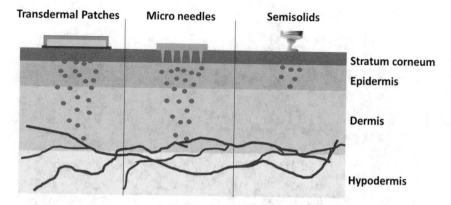

Fig. 10.2 Dermal and transdermal drug delivery from topically applied products

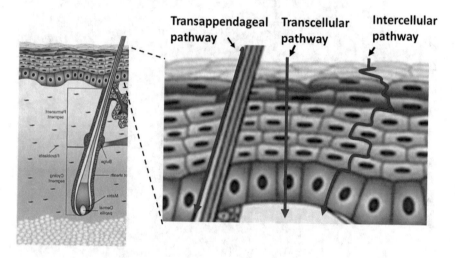

Fig. 10.3 Passive drug diffusion through three different pathways from stratum corneum (**a**), transappendageal pathway (**b**), and transcellular pathway © to intercellular pathway

through intact skin and reaches systemic circulation in a controlled manner e.g., transdermal patches and microneedles [11].

The dermal and transdermal drug delivery is illustrated in Fig. 10.2.

10.3.1 Permeation Through the Skin

Drugs' permeation through the skin is achieved by different pathways, such as intercellular pathway, transcellular pathway, and skin appendages (sweat glands, hair follicles, etc.) as shown in Fig. 10.3 [12].

The Intercellular Lipid Pathway It is the passage through the lipid matrix present in the intercellular spaces of the keratinocytes. Polar molecules utilize the free space between a lamella and the outer membrane of corneocyte [13].

The Transcellular Pathway The transcellular route passes through the skin and crosses the lipid structures of interlamellar region across the corneocytes; the corneocytes' cell membrane is highly impermeable due to which the transcellular route is considered least important although this assumption needs further investigations [14].

Through Skin Appendages In recent studies, it is considered that skin appendages provide an efficient route of drug permeation. Drug delivery is achieved across sebaceous or sweat glands and hair follicles. In the past, this route was ignored as it provides negligible contribution of skin surface [15, 16]. In topical preparations containing polymeric nanoparticles, it was reported that permeation of drug is enabled through the follicle or the follicle blockers by avoiding the penetration of drug. Therefore, besides other routes of drug delivery, the transappendageal route acts as a valuable pathway in skin permeation [17].

10.3.2 Dermal Drug Delivery (e.g., Ointments and Creams)

The product selection for dermal drug delivery depends on factors such as intended therapeutic use, availability of drug at the site of action, and ease of application. These products are typically semisolids that vary in composition and physical properties [18] and examples are shown in Table 10.1.

In this section we will particularly focus on two of the most common semisolid products used for dermal drug delivery, i.e., ointments and creams.

Traditionally, ointment is defined very generally. For instance, ointments have been defined by the USP 31 Chapter ⟨1151⟩ as "semisolid preparations intended for external application to the skin or mucous membranes." However, in pharmaceutics "ointment" dosage form is considered to provide drug incorporated into an oleaginous ointment base. Therefore, a more specific definition of ointment would be "a viscous oleaginous or polymeric semisolid dosage form (emulsions or suspensions) applied over the skin mucous membranes."

USP 31 Chapter ⟨1151⟩ defines creams as "semisolid dosage forms containing one or more drug substances dissolved or dispersed in a suitable base." According to this definition, creams can fit in the general definition of ointments. However, they are given special section in USP 31 Chapter ⟨1151⟩. A more clear and simple proposed definition for cream is "a dosage form comprised of a viscous semisolid emulsion for topical application to the skin or mucous membranes." Hence, creams could be considered as a subclass of ointments that would fall into two of the four classes of ointment bases i.e., water-containing absorption bases and water-removable bases [19].

Table 10.1 Physical attributes of common semisolid preparations

Classification	Typical attributes
Ointments	Ointments are viscous oleaginous or polymeric semisolid dosage form (emulsions or suspensions) for external application to the skin or mucous membranes.
	Commonly contain 50% hydrocarbons, PEGs, or waxes and less than 20% water.
	Ointment may form occlusive film on the skin, depending on the type of ointment base used.
Creams	Creams are "emulsion" class of the ointments. Since, they typically contain two of the four classes of ointment bases (i.e., water-containing absorption bases and water-removable bases).
	Creams cannot form occlusive film on the skin.
	Mildly greasy or nongreasy depending on hydrocarbon contents.
Pastes	Pastes are "suspension" class of the ointment, containing higher concentration of dispersed particles, i.e., 20–50% w/w (or more) in ointment bases.
	It may form occlusive film on the skin.
	It is less greasy than ointments.
Gel	Gels are semisolid suspension of cross-linked small particles or macromolecules interpenetrated by liquid.
	Depending on the interpenetrated liquid, gels may be classified as hydrogels (interpenetrated liquid is water) or organogel (interpenetrated liquid organic solvent).

Ointments and creams containing drugs (medicated) are clinically indicated to deliver drugs into the nose, eye, rectum, vagina etc. as an antiseptic, antipruritic, astringent, keratolytic, and anti-inflammatory. However, nonmedicated ointments and creams are used as emollient. Nevertheless, nonmedicated ointment comprising of oleaginous bases or anhydrous absorption base can also be used as protectant [20].

10.3.2.1   Formulation Considerations

Formulation of stable ointments and pastes requires the following factors to be considered).

Vehicle (Ointment Bases)

Ointment bases that act as carriers for the drug in formulation also control the absorption and drug delivery characteristics of the drug. Therefore, the choice of an appropriate ointment base is determined on one or more of the following factors:

 - Characteristics of the site of application (dry vs moist)
 - Desired rate of drug release from the base (determined by the drugs' solubility in the base)
 - Chemical stability of active ingredient in the ointment base (sensitivity vs resistance of drug to hydrolysis)
 - Effect of drug on viscosity of ointment base (dependent on the physical parameters and desired drug concentration and composition of the ointment base)
 - Water uptake capacity of the formulation (hydrophobic vs water-miscible)
 - Proposed use or indication of the dosage form

According to the USP 31 Chapter (1151), ointment bases may be categorized into four classes on the basis of physicochemical properties, i.e., (i) hydrocarbon bases, (ii) absorption bases, (iii) water-removable bases, and (iv) water-soluble bases, as summarized in Table 10.2.

Additives/Alternative Vehicles

Hydrophobic liquid solvents used either additionally or alternatively in hydrophobic and absorption bases include vegetable oils, organic esters, and liquid silicone.

Vegetable oils, such as arachis oil and coconut oil may be employed to enhance the emollient property of the formulation or in some instances to replace mineral oil [21].

Organic esters, such as isopropyl myristate may be added either to achieve better dissolution of the drug in ointment base or to replace mineral oil for improving the spreadability. Liquid silicone or polydimethylsiloxane may be used when its water-repellent properties are desired.

Gel formulations may also contain co-solvents like propylene glycol, glycerol, and polyethylene glycol in order to improve drug solubility in dosage form and/or to increase drug permeability. Also, in certain cases if the drug is chemically unstable in aqueous solvent alternative solvent may be used [22].

Table 10.2 Characteristics of ointment bases

Base and its characteristics	Uses	Examples
Hydrocarbon (oleaginous bases)	Protectant	*Hard paraffin*: Colorless or white solid wax (melts at 47–65 °C), mixture of straight-chain hydrocarbons (C20 to C30) obtained from petroleum
Greasy	Preferred base for hydrolyzable drugs	
Anhydrous	Emollient (restricts water loss)	*Soft paraffin (petroleum jelly)*: Contain both solid and liquid hydrocarbons (melts at ≤45 °C)
Cannot absorb water		*Liquid paraffin (also called mineral oil)*: A mixture of saturated aliphatic (C14–C18) and cyclic hydrocarbons
Insoluble in water		
Not washable with water		*Microcrystalline wax*: Mixture of cycloalkanes (naphthalene), linear alkanes, and branched alkanes (C41–C57)
Occlusive		
Absorption base	Protectant	*Hydrophilic petrolatum USP*: A mixture of petrolatum, cholesterol, stearyl alcohol, and white beeswax
(a) *Anhydrous absorption base*	Emollient	
Contain oleaginous base and w/o surfactant	Vehicle for solid and liquid drugs (also for aqueous drug solutions)	*Lanolin (wool fat) and hydrous lanolin (wool alcohol)*: Waxlike material that is derived from sheep's wool
Contain no water (anhydrous)		*Beeswax (yellow and white)*: Natural beeswax (yellow beeswax) and its beached form (white beeswax) consists of esters of aliphatic alcohols (C24–C36) and linear aliphatic fatty acids (up to C36)
Can absorb water (limited)		
Insoluble in water		
Not washable with water		
Occlusive		
(b) *Water-containing absorption base*	Emollient	*Oily cream BP*: A w/o emulsion ointment base consisting of wool alcohols (50% w/w), magnesium sulfate, phenoxyethanol, and water
Contain oleaginous base and water (<45% w/w) and w/o surfactant	Vehicle for solid and liquid drugs (also for aqueous drug solutions)	*Cold cream USP*: A w/o emulsion ointment base consisting of beeswax, spermaceti, mineral oil, borax, and water
Can absorb water (limited)		
Insoluble in water		
Not washable with water		
Occlusive		

(continued)

Table 10.2 (continued)

Base and its characteristics	Uses	Examples
Water-removable base	Emollient	*Hydrophilic ointment USP*: An o/w emulsion. The oil phase is composed of stearyl alcohol and soft paraffin, while the water phase is composed of sodium lauryl sulfate, propylene glycol, methylparaben, and propylparaben dissolved in water
Contain oleaginous base, water (>45% w/w) and o/w surfactant	Vehicle for solid and liquid drugs (also for aqueous drug solutions)	
Can absorb water (limited)		
Insoluble in water		
Washable with water		
Nonocclusive		
Water-soluble base	Emollient	*Polyethylene glycols (PEGs)*: Lower molecular weight PEGs (<1000) are colorless viscous liquids, whereas PEGs of higher molecular weight are waxy solids. A blend of different molecular weights is used as a base to attain desired consistency of ointment
Does not contain oleaginous base (lipid free)	Vehicle for solid and liquid drugs (also for aqueous drug solutions)	
Can absorb water (limited)		
Soluble in water		
Washable with water		
Nonocclusive		

Preservatives

Addition of antimicrobial preservatives in semisolids or O/W emulsions is essential to prevent and resist any microbial attack that may affect the physicochemical properties of the formulation or may also cause toxic effects.

Two or more preservatives may be added in a formulation to broaden the antimicrobial spectrum while keeping the compatibility problems in consideration.

In order to maintain the minimum inhibitory concentration of preservative in aqueous phase of the formulation, which otherwise may be altered owing to its partitioning between oil and aqueous phases, it is important to consider pH of the formulation to ensure that preservative retains its undissociated form (active form). It may also be compensated by using high initial concentration of the preservative [23].

Examples of preservatives used in semisolid preparations are as given in the Table 10.3.

Antioxidants and Humectants

Antioxidants are employed in semisolid preparations (particularly ointments) to prevent the auto-oxidation of oils, emulsifiers, and the active ingredient itself during the shelf life that may have effect on the stability and physicochemical properties of the product.

Table 10.3 Common excipients used in ointments and creams

Excipients	Examples
Vehicle	Ointment bases (detailed in Table 10.1)
Additives/alternative vehicles	*Liquid silicone or polydimethylsiloxane*
	Vegetable oils: Coconut oil and arachis oil
	Organic esters: Isopropyl myristate
Emulsifying agents	*Anionic*: Sodium oleate, calcium stearate
	Cationic: Cetrimide
	Amphoteric: Phospholipids
	Nonionic: Sorbian esters (called Span series), polyoxyethylene fatty acid derivatives of the sorbitan esters (called Tween series)
	(for detail description see table)
Antimicrobial agents	Phenol, chlorocresol, benzyl alcohol, methyl and propyl para-hydroxybenzoic acid, benzoic acid, etc.
Antioxidants	*Hydrophilic antioxidants*: Sodium sulfite, sodium metabisulfite, ethylenediaminetetraacetic acid (EDTA), ascorbic acid, etc.
	Hydrophobic antioxidants: Butylated hydroxyanisole (BHA)
	Butylated hydroxytoluene (BHT)
Humectants	Glycerol, propylene glycol, and sorbitol
Penetration enhancers	*Short-chain alcohols*: Ethanol, isopropyl alcohol
	Long-chain alcohols: Hexanol, myristyl alcohol, oleyl alcohol
	Sulfoxides: Dimethyl sulfoxide, decylmethyl sulfoxide
	Terpenes: Eugenol, menthol, D-limonene, etc.
	Surfactants: Phospholipids, sodium lauryl sulfate, etc.
	Glycols: Propylene glycol, dipropylene glycol, etc.

Hydrophilic or hydrophobic antioxidants may be chosen depending upon the nature of vehicle being used in the formulation.

Humectants are also added in semisolid preparations (particularly gels) for the purpose of preventing or minimizing the evaporation of water from the formulation during shelf life and application. They have been proposed to prevent loss of water from the surface of skin, thereby enhancing the penetration and solubility of drug as well as hydration of skin [21]. Commonly used humectants are as given in Table 10.3.

Penetration Enhancers

Delivery of drug to its target site depends largely on the penetration of semisolid formulation through the skin. This rate-limiting process of percutaneous absorption follows a progression of steps i.e., release of drug from the formulation and its diffusion onto skin surface, the drug partitioning into and diffusion through the stratum corneum followed by diffusion into other epidermal layers and then finally partitioning into the fat deposits and blood vessels present in dermis [24].

Chemical penetration enhancers function to reversibly reduce and overcome the primary barrier resistance offered by stratum corneum thereby allowing sufficient concentrations of drug to be penetrated into the skin tissues and systemic circulation [25].

Examples of polymeric chemical penetration enhancers are as given in Table 10.3.

10.3.2.2 Preparation of Ointments and Creams

Lab-Scale Manufacturing

Preparation procedures of ointment should ensure absolute uniformity of final product free from grittiness and phase separation of the aqueous and oil phases.

For laboratory-scale and extemporaneous compounding ointments are commonly prepared by two techniques.

Fusion Method

It involves melting together all or some ingredients followed by stirring till a homogeneous mixture is obtained. This technique is favored for formulation containing hard fats and/or waxes. Moreover, the solid ingredients of the formulation are soluble in the base. Fusion method is typically employed for incorporation of large quantities of hydrophobic components [26].

Direct Incorporation Method

This method involves mixing of the solid components with a fraction of base by trituration or levigation (wet grinding with aid of levigating agent) followed by stepwise dilution with the base until a uniform preparation is obtained. The method is used when the base contains liquid components (soft fats and/or oils), and the solid components of the formulation are insoluble in the base. Liquids and finely divided insoluble powders are incorporated into base by means of trituration, while insoluble coarse powders are incorporated into base by means of levigation [27].

Large-Scale Manufacturing

Industrial manufacturing of ointments and creams follows an easy and straightforward procedure, somewhat similar to the one described for industrial manufacturing of emulsions in Chap. 5.

For manufacturing of ointments, drug powder is dispersed into preheated ointment base. Heating lowers the viscosity of the base, thus improving mixing of the ingredients with the base.

Manufacturing of creams involves incorporation of drug as distinct liquid phase into the vehicle. The hydrophilic and hydrophobic components are dissolved in water and oil phases respectively with the help of stirring and heating. Subsequently, the water and oil phases (usually maintained at 70 °C) are mixed and homogenized. For example, the formulation of gentamicin sulfate cream is given in Table 10.4.

Preparation of Oil Phase: White soft paraffin and liquid paraffin are loaded and melted at 70 °C in a stainless-steel vessel.

Table 10.4 Formulation of gentamicin sulfate cream

Serial no.	Material name	Quantity (kg or g)
1.	Gentamicin sulfate	1.82
2.	White soft paraffin	150
3.	Cetomacrogol 1000	18
4.	Cetostearyl alcohol	72
5.	Chlorocresol	1.0
6.	Liquid paraffin	60
7.	Monobasic sodium phosphate	3.0
8.	Purified water	694.17

Preparation of Aqueous Phase: 608 g of water is charged in a stainless-steel vessel and heated up to 70 °C.

Preparation of Cream (O/W): The molten oil phase is transferred into aqueous phase at 70 °C through a stainless-steel filter and simultaneously mixing it slowly. The crude cream is passed through homogenizer for 10–15 min while temperature is maintained at around 70 °C.

Incorporation of Drug in the Cream: Gentamicin is dissolved in in 86.17 g of water while mixing at 50 °C. The drug solution is transferred into cream while stirring at 50 °C.

Homogenization: The final cream product is passed and circulated through homogenizer for 10–15 min. The temperature is lowered to 25 °C during homogenization. When the cream temperature reaches 25 °C, it is ready for packaging.

10.3.2.3 Tests for Evaluation of Semisolids

Evaluation of the quality attributes by various compendial and non-compendial tests is required to assure good quality and avoid batch-to-batch variation.

Appearance

Providing qualitative description of the semisolid product is required to ascertain if the product meets the acceptance criteria for appearance of finished dosage form and packaging as specified and claimed by the manufacturer on the label. Any sort of specific changes in the color and consistency such as discoloration, crystallization, separation, shrinkage, grittiness, etc. should be identified by visual evaluation [28].

Uniformity of Dosage Units

The extent of uniformity of the API amount among the dosage units of a semisolid product should be determined to ascertain the consistency of dosage units. Both the content uniformity and weight variation methods can be used for this test and are discussed in the USP chapter, titled "Uniformity of Dosage Units" [29].

Water Content

Depending on the individual formulation, water content test ought to be performed when considered appropriate and is discussed in USP chapter 921, titled "Water Determination."

Limits for Presence of Microbes

Unsterilized aqueous dermatological products should be assessed for presence and/or amount of the microbes that have potential to cause skin infection and reduce the pharmacological activity, such as *Escherichia coli, Staphylococcus aureus, Pseudomonas aeruginosa,* and various *Salmonella* species. Additionally, the semisolid preparations intended for application at urethral, vaginal, and rectal sites should be screened for presence of yeasts and molds [30].

Procedures for microbial tests should be performed according to methods demonstrated in USP chapter 61 titled "Microbiological Examination of Nonsterile Products: Microbial Enumeration Tests" and 62 titled "Microbiological Examination of Nonsterile Products: Tests for Specified Microorganisms." The acceptance criteria for microbial count are given in USP chapter 1111.

Preservative Content

Standard concentration of the antimicrobial preservatives in semisolid products should be demonstrated. Acceptance criteria is determined by the range of preservative content required for maintaining the microbial quality of drug during its shelf life and usage.

Method and acceptance criteria for assessing the effectiveness of preservative are demonstrated in the USP chapter 51.

Antioxidant Content

The tests for content of antioxidants should be determined if the drug product contains antioxidants and the oxidative degradation cannot be detected by impurity test. Standard concentration of antioxidants in semisolid products should be established.

Acceptance criteria are determined by the range of antioxidant content required for maintaining the stability of drug product during its shelf life and usage [31].

Sterility

Sterility of semisolid products that need to be sterile (according to pharmacopoeia) such as ophthalmic ointments and products intended for application on burns or open wounds should be ensured by sterility tests, such as the membrane filtration method and direct inoculation test. Details of the test are given in the USP chapter 71.

pH

Semisolid products containing adequate amount of water or aqueous phase should be tested for pH values before release of a batch for ensuring quality consistency among various batches. The test is not included in pharmacopeial drug product monograph; rather it is formulation dependent and specified by the manufacturer of the drug product [19].

Particle Size

Semisolid drug products should be tested for particle size, shape, aggregation, or crystal habit of the particles that have possibly occurred during the processing or storage of these products. Such alterations may lead to the compromised quality of the product. The test should be performed before release of a batch as well as at specified stability time points for monitoring. The test is formulation dependent and specified by the manufacturer of the drug product [32].

Apparent Viscosity

Rheological properties of the semisolid preparations should be tested by determination of the apparent viscosity. Details of the viscosity measurement procedure are discussed in the USP chapter 911. The test is formulation dependent and specified in individual monographs.

Uniformity in Containers

Uniformity of the semisolid drug product within packaging tubes and containers should be evaluated as the product may have encountered phase separation or deviation in physical appearance during manufacturing or storage. Thus, uniformity of the dosage form should be evaluated before releasing a batch and during the shelf life for validation of product integrity. The procedure and acceptance criteria are elaborated in the USP chapter 3.

10.3.3 Transdermal Patches

When drug is delivered through the skin to reach the systemic circulation, it is termed as *transdermal drug delivery (TDD)*. Transdermal drug delivery system is used when systemic effect is desired. TDD is a safe, painless, and noninvasive method of drug transport and it offers great advantage over other conventional delivery routes like oral, parenteral, and hypodermal shots. TDDS uses skin as drug administration site. Drug is delivered by the aid of adhesive patches or other transdermal devices like microneedles [33, 34].

A transdermal or skin patch is medicated adhesive patch which essentially delivers a specific dose of a medication to the circulation after passing through the skin when placed on skin. Transdermal patches basically use a special membrane to control the rate at which drug contained within the patch can pass through the skin and into bloodstream in a controlled manner [35]. The patches were developed for the first time in the 1970s, and the first patch approved by FDA was scopolamine which was used for treatment of motion sickness. Transdermal patches applied on the skin eliminate the need for vascular access by syringe or the use of pumps.

10.3.3.1 Basic Components of Transdermal Patches

Polymer Matrix or Drug Reservoir

The polymer matrix or drug reservoir is the primary component used to control the release of drug from patch. It should be nonreactive with active agent, biodegradable, nontoxic, stable, and facilitate the uniform release of drug throughout.

The Drug

The drug to be incorporated in transdermal patch should be potent and non-irritating and have short half-life, with low melting point and low molecular weight ($<$400). The drug in the patch should follow zero order kinetics of flow with broad therapeutic index and low-dose requirements.

Permeation Enhancers

Permeation enhancers are used to promote the sorption of drug from drug delivery system onto the skin. Various solvents (methanol, ethanol, propylene glycol, isopropyl palmitate), anionic surfactants (sodium lauryl sulfate, dioctyl sulfosuccinate), nonionic surfactants (Pluronic F127 and F68), and bile salts are used as permeation enhancers.

Pressure-Sensitive Adhesive

Pressure-sensitive adhesive is used for fastening the patch with the skin. The ideal characteristics are that it should adhere with the skin without disturbing the normal flora of the skin and retained on the skin while bathing or exercise. It should not leave an unwashable residue and can be removed easily from the skin.

Backing Laminates

Backing laminates are impermeable and flexible and prevent the drug to release from the top of the patch. In transdermal patches, plastic backing with absorbent pad, adhesive foam pad, and metallic plastic laminates are used.

Release Liner

Release liners are basically part of primary packaging material which avoids the drug loss due to the escape of drug into the adhesive layer in the duration of storage. It is composed of occlusive (polyethylene), nonocclusive (paper fabric), and a release coating layer (Teflon or silicon) [36, 37]

10.3.3.2 Types of Transdermal Patches

Single-Layer Drug in Adhesive

In these patches, drug is contained in the adhesive layer. In these patches, adhesive layer is responsible to adhere several layers together along with the skin as well as to release the drug in a controlled manner [38].

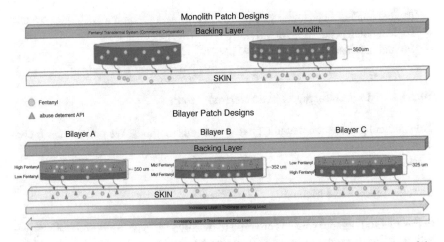

Fig. 10.4 Figure 10.4 should be placed under 10.3.3.2 section Graphical representation of the different spatial distributions and layer thicknesses of the prototypes used to control multidrug release [40]

Multilayer Drug in Adhesive

Basically, these are like single-layer system but separated by a membrane. In these patches, one layer is designed for immediate release of drug while other layer is for control release of drug from reservoir.

Reservoir Reservoir is considered as a separate layer of drug in liquid compartment containing drug solution and suspension that are separated by the adhesive layer.

Matrix The matrix system in these patches contains a layer of semisolid matrix in which drug solution or suspension is present.

Vapor Patch In these patches, adhesive layer supports to combine various layers together and release vapors as well. They basically provide essential oils and are also used in conditions of decongestion and to aid in smoking cessation [39].

10.3.3.3 Advantages of Transdermal Patches

- Topical patches are painless and noninvasive method to deliver substances directly into body.
- Topical patches are best suitable for drugs that are broken down by stomach acids and have extensive first-pass effect.
- Topical patches deliver the drug in steady and controlled manner over longer period of time.

- Topical patches have lesser side effects as compared to oral medications.
- Topical patches are user-friendly.
- Topical patches are economical [41].

10.3.3.4 Manufacturing of Transdermal Patches

Transdermal patches are manufactured by various methods. We will discuss few of them here.

Circular Teflon Mold Method

In this method organic solvents are used to dissolve various proportions of polymer and drug. Plasticizers are added in the same drug polymer solution. Similarly, permeation enhancers are also added in the remaining half portion of the organic solvent and then incorporated in the drug polymer solution. Afterward, the overall solution is stirred for ~12 h following addition into a circular Teflon mold which has been enclosed with an inverted funnel to regulate solvent vaporization placed in a laminar flow hood (LFH) on a levelled surface for 24 h. The dried films obtained were kept for 24 h at 25 ± 0.5 °C in a desiccator before evaluation [5].

Asymmetric TPX Membrane Method

A transdermal patch can be prepared with heat-sealed polyester film supported by a backing membrane having concave diameter of approximately 1.0 cm. The drug is dispensed into the concave membrane covered with TPX asymmetric membrane and sealed by an adhesive. TPX {poly(4-methyl-1-pentene)} asymmetric membrane can be fabricated by utilizing dry/wet inversion process. A polymer solution is prepared by dissolving TPX in a mixture containing solvent (such as cyclohexane) and non-solvent additives at 60 °C temperature. The polymer solution is then kept for 24 h at 40 °C followed by casting on a glass plate with a predetermined thickness control by using a Gardner knife. A film is formed by evaporating the organic solvent at 50 °C for 30 s and immediately immersed in a coagulation bath already maintained at 25 °C temperature. Ten minutes after the immersion, the membrane is removed and air-dried in a circulation oven for 12 h at 50 °C. The release of drug from the TPX asymmetric membrane is dependent on the structure of membrane which can be fabricated by varying the non-solvents in casting solution [42].

EVAC Membrane Method

Ethylene vinyl acetate copolymer (EVAC) membranes are utilized as control-release transdermal patches. One percent of Carbopol reservoir gel, polyethylene (PE), or in case of water-insoluble drug propylene glycol is the choice of EVAC. Drug dissolution in PE or PG is carried out, and Carbopol gel resin will be incorporated and neutralized with 5% solution of sodium hydroxide. The drug in the gel is employed over the backing layer and rate controlling membrane will be placed over the drug containing gel and hermetically sealed by heat to avoid leakage [43].

Some other methods like aluminum-backed adhesive, IPM membrane, mercury substrate, proliposome, and free film can be used for the manufacturing of transdermal patches.

10.3.3.5   Quality Control Parameters

Physical Appearance

Physical appearance of patches is checked visually for color, clarity, and surface texture.

Weight Variation

Weight variation is checked by cutting 2×2 cm^2 from the film and weight on electronic balance.

Thickness

Thickness of patch is measured at three different places with the help of vernier caliper and average thickness is calculated.

Tensile Strength and % Elongation

The tensile strength at breaking point of film is measured with the help of a pulley base tensile strength apparatus:

$$Tensile\,strength = \frac{tensile\,load\,at\,break}{Cross\,sectional\,area}$$

$$(10.1)$$

The % elongation is calculated by noting the length just before the breakpoint by the following equation [44]:

$$\%\text{Elongation} = \frac{\left[\text{Final length} - \text{initial length}\right]}{\text{initial length}} \times 100$$

(10.2)

Folding Endurance

Folding endurance can be measured by repeatedly folding a 2 cm^2 strip at a same point until it breaks. The value of folding endurance is determined by the number of times the film is folded at same point without being broken.

Percent Moisture Loss

Percent moisture loss is calculated by weighing individual patch and keeping them in a calcium chloride containing dessicator. The final weight is noted when no further change is observed in weight. The % moisture loss is calculated with the help of the following equation [45]:

$$\text{Moisture content}\% = \frac{\text{Initial weight} - \text{Final weight}}{\text{Initial weight}} \times 100$$

(10.3)

Percent Moisture Uptake

Percent moisture uptake is calculated by weighing the individual patch and placing them in a desiccator and humidity is maintained 80–90% RH with the help of saturated solution of ammonium chloride. The patches are kept until no more change is observed in weight. Percent moisture uptake is calculated with the help of the following equation [46]:

$$\text{Moisture uptake}\% = \frac{\text{Initial weight} - \text{Final weight}}{\text{Initial weight}} \times 100$$

(10.4)

Drug Content Uniformity

Drug content is calculated by cutting the patch 2 cm^2 and dissolved in phosphate buffer solution on a magnetic stirrer for 24 h, and drug content is determined on UV-visible spectrophotometer or HPLC.

10.3.3.6   Therapeutic Applications of Transdermal Patches

Mostly sold patches in the USA is nicotine patch, which releases nicotine in a controlled manner to assist in cessation of tobacco. Two opioid medications that provide relief from severe pain are fentanyl (Duragesic) and buprenorphine (BuTrans). Estrogen patches are used to control menopausal symptoms. Nitroglycerin patches are used for angina.

10.3.4   Transdermal Microneedles

Basically, microneedles are of micron-sized needles which can penetrate easily through epidermal and dermal layer without pain.

Microneedles' height ranges from 10 to 200 micrometer and a width of 10–50 micrometers. Microneedles are widely used in TDDS because they are safe, efficient, painless, and convenient. Microneedles are the rationalized way to deliver the drug through the skin. Through microneedles large molecules can be administered. In contrast to the hypodermic needles, microneedles facilitate faster healing of injection sites and lower microbial penetration. First-pass metabolism is also avoided [47, 48].

10.3.4.1   Types of Microneedles

Microneedles are made up of insoluble metal alloys (such as titanium, stainless steel, nickel iron), insoluble silicon, and biodegradable polymers such as polyvinylpyrrolidone (PVP), poly(lactic-co-glycolic acid) (PLGA), carboxymethyl cellulose (CMC), hyaluronic acid, etc.

Based on the design and drug delivery principle, microneedles are divided into four classes.

Solid Microneedles

Solid microneedles basically create microchannel in stratum corneum. It generally increases the permeability by creating microchannels in the skin, and the drug is rubbed over that area or the needles are coated with drug. The microneedles use passive diffusion pathway [49].

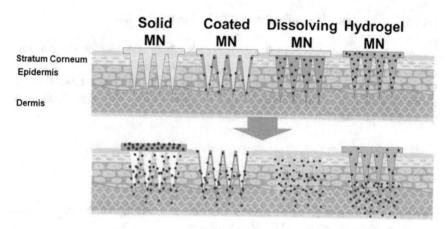

Fig. 10.5 Schematic representation of different types of microneedles and their drug delivery methods [53]

Hollow Microneedles

Hollow microneedles generally contain hollow bore channels, which allows the passage of drug through it. After insertion of needle into the skin, the drug passes through the bore, after which it diffuses into systemic circulation [50].

Dissolving Microneedles

The drug in dissolving microneedles is encapsulated in a suitable biodegradable polymer, which are inserted in the skin for drug release [51].

Coated Microneedles

Coated microneedles are basically solid microneedles onto which drug is coated. When it is inserted into the skin, the drug gets dispersed into systemic circulation [52].

10.3.4.2 Preparation of Microneedles

Microneedles can be manufactured by the following method divided into three parts utilizing microelectrochemical system (MEMS).

The first step is deposition in which a thin film having a thickness up to 100 micrometers is prepared. The second step is patterning which is performed on the film using lithography. Various types of lithography techniques can be used for this purpose like photolithography, ion beam lithography or X-ray lithography, and electron beam lithography. In some cases, diamond patterning is also used for lithography. The next step in fabrication of microneedle is etching. In etching process, the unprotected parts of material's surface are cut to form a design with the help of a strong acid or mordant. Etching is divided into two types: dry etching and wet etching. The selection of etching depends upon the material used and type of microneedle [49, 54].

References

1. Desai, P., R.R. Patlolla, and M. Singh, *Interaction of nanoparticles and cell-penetrating peptides with skin for transdermal drug delivery.* Molecular membrane biology, 2010. **27**(7): p. 247–259.
2. Ham, A.S. and R.W. Buckheit Jr, *Current and emerging formulation strategies for the effective transdermal delivery of HIV inhibitors.* Therapeutic delivery, 2015. **6**(2): p. 217–229.
3. Kim, H.S., et al., *Advanced drug delivery systems and artificial skin grafts for skin wound healing.* Advanced drug delivery reviews, 2019. **146**: p. 209–239.
4. Malvey, S., J.V. Rao, and K.M. Arumugam, *Transdermal drug delivery system: A mini review.* Pharma Innov. J, 2019. **8**: p. 181–197.
5. Benson, H.A., et al., *Topical and transdermal drug delivery: From simple potions to smart technologies.* Current drug delivery, 2019. **16**(5): p. 444–460.
6. Williams, A.C., *Topical and transdermal drug delivery.* Aulton's Pharmaceutics: The Design and Manufacture of Medicines. 5th ed. Edinburgh: Elsevier, 2018: p. 715–725.
7. Wong, R., et al., *The dynamic anatomy and patterning of skin.* Experimental dermatology, 2016. **25**(2): p. 92–98.
8. Gilaberte, Y., et al., *Anatomy and Function of the Skin,* in *Nanoscience in Dermatology.* 2016, Elsevier. p. 1–14.
9. Losquadro, W.D., *Anatomy of the skin and the pathogenesis of nonmelanoma skin cancer.* Facial Plastic Surgery Clinics, 2017. **25**(3): p. 283–289.
10. Haney, B., *Aesthetic Procedures: Nurse Practitioner's Guide to Cosmetic Dermatology.* 2020: Springer.
11. Roxhed, N., et al., *Painless drug delivery through microneedle-based transdermal patches featuring active infusion.* IEEE Transactions on Biomedical Engineering, 2008. **55**(3): p. 1063–1071.
12. Hadgraft, J. and R.H. Guy, *Feasibility assessment in topical and transdermal delivery: mathematical models and in vitro studies,* in *Transdermal drug delivery.* 2002, CRC Press. p. 1–23.
13. Alexander, A., et al., *Approaches for breaking the barriers of drug permeation through transdermal drug delivery.* Journal of Controlled Release, 2012. **164**(1): p. 26–40.
14. Ng, K.W. and W.M. Lau, *Skin deep: the basics of human skin structure and drug penetration,* in *Percutaneous penetration enhancers chemical methods in penetration enhancement.* 2015, Springer. p. 3–11.
15. Hadgraft, J., *Skin, the final frontier.* International journal of pharmaceutics, 2001. **224**(1–2): p. 1–18.
16. Lademann, J., et al., *Hair follicles–an efficient storage and penetration pathway for topically applied substances.* Skin pharmacology and physiology, 2008. **21**(3): p. 150–155.
17. Neubert, N., et al., *The molybdenum isotopic composition in river water: constraints from small catchments.* Earth and planetary science letters, 2011. **304**(1–2): p. 180–190.
18. Olejnik, A., J. Goscianska, and I. Nowak, *Active compounds release from semisolid dosage forms.* Journal of pharmaceutical sciences, 2012. **101**(11): p. 4032–4045.
19. Bora, A., S. Deshmukh, and K. Swain, *Recent advances in semisolid dosage form.* International Journal of Pharmaceutical Sciences and Research, 2014. **5**(9): p. 3596.

20. Gupta, P. and S. Garg, *Semisolid dosage forms for dermatological application.* Pharmaceutical technology, 2002: p. 144–62.

21. Garg, T., G. Rath, and A.K. Goyal, *Comprehensive review on additives of topical dosage forms for drug delivery.* Drug delivery, 2015. **22**(8): p. 969–987.

22. Bharat, P., et al., *A review: Novel advances in semisolid dosage forms & patented technology in semisolid dosage forms.* Int. J. PharmTech Res, 2011. **3**: p. 420–430.

23. Maqbool, M.A., et al., *Semisolid dosage forms manufacturing: Tools, critical process parameters, strategies, optimization, and recent advances.* Ind. Am. J. Pharm. Res, 2017. **7**: p. 882–893.

24. Thakur, V., B. Prashar, and S. Arora, *Formulation and in vitro Evaluation of Gel for Topical Delivery of Antifungal Agent Fluconazole Using Different Penetration Enhancers.* Drug Invention Today, 2012. **4**(8).

25. Sadozai, S.K., et al., *Ketoconazole-loaded PLGA nanoparticles and their synergism against Candida albicans when combined with silver nanoparticles.* Journal of Drug Delivery Science and Technology, 2020. **56**: p. 101574.

26. Jabbari, A. and K. Abrinia, *A metal additive manufacturing method: semi-solid metal extrusion and deposition.* The International Journal of Advanced Manufacturing Technology, 2018. **94**(9): p. 3819–3828.

27. Khullar, R., et al., *Emulgels: a surrogate approach for topically used hydrophobic drugs.* Int J Pharm Bio Sci, 2011. **1**(3): p. 117–28.

28. Goebel, K., et al., *In vitro release of diclofenac diethylamine from gels: evaluation of generic semisolid drug products in Brazil.* Brazilian journal of pharmaceutical sciences, 2013. **49**(2): p. 211–219.

29. Siska, B., et al., *Contribution to the rheological testing of pharmaceutical semisolids.* Pharmaceutical development and technology, 2019. **24**(1): p. 80–88.

30. Mahalingam, R., X. Li, and B.R. Jasti, *Semisolid dosages: ointments, creams, and gels.* Pharmaceutical Sciences Encyclopedia: Drug Discovery, Development, and Manufacturing, 2010: p. 1–46.

31. Rasidek, N.A.M., M.F.M. Nordin, and K. Shameli, *Formulation and evaluation of semisolid jelly produced by Musa acuminata Colla (AAA Group) peels.* Asian Pacific journal of tropical biomedicine, 2016. **6**(1): p. 55–59.

32. Alves, P., A. Pohlmann, and S. Guterres, *Semisolid topical formulations containing nimesulide-loaded nanocapsules, nanospheres or nanoemulsion: development and rheological characterization.* Die Pharmazie-An International Journal of Pharmaceutical Sciences, 2005. **60**(12): p. 900–904.

33. Pastore, M.N., et al., *Transdermal patches: history, development and pharmacology.* British journal of pharmacology, 2015. **172**(9): p. 2179–2209.

34. Jamakandi, V., et al., *Formulation, characterization, and evaluation of matrix-type transdermal patches of a model antihypertensive drug.* Asian Journal of Pharmaceutics (AJP): Free full text articles from Asian J Pharm, 2014. **3**(1).

35. Isaac, M. and C. Holvey, *Transdermal patches: the emerging mode of drug delivery system in psychiatry.* Therapeutic advances in psychopharmacology, 2012. **2**(6): p. 255–263.

36. Premjeet, S., et al., *Transdermal drug delivery system (patches), applications in present scenario.* Int J Res Pharm Chem, 2011. **1**(4): p. 1139–1151.

37. Shi, Y., et al., *Design and in vitro evaluation of transdermal patches based on ibuprofen-loaded electrospun fiber mats.* Journal of Materials Science: Materials in Medicine, 2013. **24**(2): p. 333–341.

38. Padula, C., et al., *Single-layer transdermal film containing lidocaine: modulation of drug release.* European journal of pharmaceutics and biopharmaceutics, 2007. **66**(3): p. 422–428.

39. Su, X., et al., *Layer-by-layer-assembled multilayer films for transcutaneous drug and vaccine delivery.* Acs Nano, 2009. **3**(11): p. 3719–3729.

40. Davis, D.A., et al., *Complex Drug Delivery Systems: Controlling Transdermal Permeation Rates with Multiple Active Pharmaceutical Ingredients.* AAPS PharmSciTech, 2020. **21**: p. 1–11.

41. Stanos, S.P. and K.E. Galluzzi, *Topical therapies in the management of chronic pain.* Postgrad Med, 2013. **125**(4 Suppl 1): p. 25–33.

42. Mi, F.L., et al., *Fabrication and characterization of a sponge-like asymmetric chitosan membrane as a wound dressing.* Biomaterials, 2001. **22**(2): p. 165–73.

43. Ocak, F.G. and İ. Ağabeyoğlu, *Development of a membrane-controlled transdermal therapeutic system containing isosorbide dinitrate.* International Journal of Pharmaceutics, 1999. **180**(2): p. 177–183.

44. Prajapati, S.T., C.G. Patel, and C.N. Patel, *Formulation and evaluation of transdermal patch of repaglinide.* ISRN Pharm, 2011. **2011**: p. 651909.

45. Panchaxari, D.M., et al., *Design and characterization of diclofenac diethylamine transdermal patch using silicone and acrylic adhesives combination.* Daru, 2013. **21**(1): p. 6.

46. Limpongsa, E. and K. Umprayn, *Preparation and evaluation of diltiazem hydrochloride diffusion-controlled transdermal delivery system.* AAPS PharmSciTech, 2008. **9**(2): p. 464–70.

47. Hao, Y., et al., *Microneedles-Based Transdermal Drug Delivery Systems: A Review.* J Biomed Nanotechnol, 2017. **13**(12): p. 1581–1597.

48. Waghule, T., et al., *Microneedles: A smart approach and increasing potential for transdermal drug delivery system.* Biomed Pharmacother, 2019. **109**: p. 1249–1258.

49. Bariya, S.H., et al., *Microneedles: an emerging transdermal drug delivery system.* J Pharm Pharmacol, 2012. **64**(1): p. 11–29.

50. Ahmed Saeed Al-Japairai, K., et al., *Current trends in polymer microneedle for transdermal drug delivery.* Int J Pharm, 2020. **587**: p. 119673.

51. Leone, M., et al., *Dissolving Microneedle Patches for Dermal Vaccination.* Pharm Res, 2017. **34**(11): p. 2223–2240.

52. Shakya, A.K., et al., *Microneedles coated with peanut allergen enable desensitization of peanut sensitized mice.* J Control Release, 2019. **314**: p. 38–47.

53. Jung, J.H. and S.G. Jin, *Microneedle for transdermal drug delivery: current trends and fabrication.* J Pharm Investig, 2021: p. 1–15.

54. Bariya, S.H., et al., *Microneedles: an emerging transdermal drug delivery system.* Journal of Pharmacy and Pharmacology, 2012. **64**(1): p. 11–29.

Chapter 11
Sterile Products

Muhammad Naseer Abbas, Waqar Iqbal, and Shahzeb Khan

Abstract Sterile products include parenterals, surgical dressings, sutures, ligatures, and ophthalmic preparations. Parenteral products are meant for delivering the drug beside the intestine, and hence bioavailability of drugs may be enhanced. This chapter covers various types of parenterals, their preparation methods, and their advantages over other route of administration. Moreover, ophthalmic preparations have been also discussed in detail along with their different types and their methods of preparation. Moreover, various excipients used in formulation of parenterals and ophthalmic products have been also discussed, and their impact on products stability and efficacy have been also elaborated. Moreover, the chapter also enfolds a detailed description on key environmental parameters needed to be controlled for sterile pharmaceutical products.

Keywords Parenterals · Ophthalmic products · Ocular inserts · Environmental monitoring · Clean room

11.1 Introduction

Many pharmaceutical products such as injections, dressings, sutures etc., that come in contact with broken skin, bloodstream, or internal organs must be sterile in order to avoid the possibility of microbial infection. Pathogenic microorganisms are obviously primary threat to sterility; however, nonpathogenic microorganism if

M. N. Abbas (✉) · W. Iqbal
Department of Pharmacy, Kohat University of Science and Technology, Kohat, Pakistan
e-mail: naseerabbas@kust.edu.pk

S. Khan
University of Malakand, Chakdara, Pakistan

accidentally gains access to body cavities in a large number may pose danger to patients. Therefore, injections, dressings, sutures, ligatures, ophthalmic preparations, irrigation fluids, implants, etc. must be sterile [1].

11.2   Parenteral Dosage Forms

Dosage form purpose is to deliver drug to site of action, i.e., whether it is through oral route in the form of tablets, capsules, syrups, etc. or through parenteral route. The word parenteral is from two Greek words, *para* means beside and *enteron* means intestine. So, parenteral delivers the drug beside the intestine. In parenteral dosage form, we breech the barrier (i.e., the skin which is one of the tough barriers of human body) with the help of a needle [2]. Different routes of parenteral administration are summarized in Table 11.1.

Advantages Parenteral preparations are adventitious as they provide the rapid onset of action because drug is directly entering into the systemic circulation, and

Table 11.1 Common routes of parenteral administration [2]

Routes of administration	Important points
Intravenous route	Injection into the vein
	Quickest onset and avoids irritation of tissues
	Usually, for small volume parenterals (SVP) as well as large volume parenterals (LVP)
	Pharmacological response can hardly be reversed
	Requires skilled personnel
Intramuscular route	Injection to the gluteal muscle, thighs, deltoid (upper arm)
	Usually, for SVP up to 2 mL or some time up to 4 mL
	Can be used for sustained release (injectable depot)
Subcutaneous route	Injection into the subcutaneous layer (fat beneath the skin)
	Usually, for SVP up to 2 mL (e.g., vaccines, insulin, etc.)
	In case of emergencies when veins are not traceable LVP, 250–1000 mL can be administered (hypodermoclysis)
Intradermal route	Injection into dermal layer of the skin, for diagnostic purpose, e.g., test for allergies
Intra-arterial route	Injection into an accessible artery, for example, in drug addicts whose veins are difficult to be found
Intrathecal route	Injection into the cerebrospinal fluid in order to bypass blood-brain barrier
Intradural and extradural	Injection into the spinal cord, within the dural membrane (intradural) or outside the dural membrane (extradural), e.g., for spinal anesthesia
Intracardiac injection	Injection directly into the muscles of the heart, e.g., administration of epinephrine in emergency case of cardiac arrest when vessels are not easily accessible

secondly parenteral provides predictable effect. The bioavailability is almost complete or 100% because parenteral bypasses the first-pass effect. There is no or less absorption problems for parenteral because this route avoids GI track and moreover provides a reliable drug administration to very ill or unconscious patient [2].

Disadvantages This route involves irreversible drug administration and bears more risks as compared to others. This invasive route can cause fear, pain, tissue damage, or infection. Quite riskier mode of administration and preparation must be sterile. Without the presence of expert, parenteral route of drug administration becomes difficult [2].

11.2.1 Brief History of Parenterals

In the mid-nineteenth century, parenteral dosage forms were introduced. The first official injection was morphine, which came in British Pharmacopeia in 1867, and then cocaine injection in 1898. In national formulary the first injection was introduced in 1926, while the first monograph for injection appeared in USP in 1942. Now the current USP contains more than 500 injection monographs. Injectable dosage form is increasing rapidly. Most of the new products that come in market nowadays are in the form of injectables. It is because most of the new drugs have bioavailability problems, i.e., if we deliver the drug through oral route by means of any dosage form like tablet, capsule, syrups, etc., it has a very limited absorption from GIT. Therefore for enhancement of bioavailability, injectables are preferred [3].

11.2.2 Types of Parenterals

11.2.2.1 Parenteral Solutions

Most of the products are in solution form. Drug solutions suitable for parenteral administration are known as injections/injectables. Injectable solutions are prepared by dissolving drug (active ingredients) and excipients, maintaining a specific pH. After this solution is filtered through 0.22 μm membrane filter and finally autoclaved (if thermostable) [3].

Most of the solutions have similar viscosity and surface tension to that of water, while some solutions like streptomycin injections and ascorbic acid injections are quite viscous. Large and small volume parenterals containing no antimicrobial agents should be terminally sterilized. Volume of parenteral should be greater than the mentioned label because some of the content may retain within the container [4].

11.2.2.2 Parenteral Suspension

Formulation of parenteral suspensions is quite difficult. An injectable suspension contains various components like it has active ingredients that are suspended in antimicrobial preservative aqueous vehicle, surfactant, dispersing or suspending agent, and a buffer or a salt [4].

11.2.2.3 Parenteral Emulsion

Emulsions are dispersion of one immiscible liquid into another liquid with the help of emulsifying agents. This system is made by the addition of emulsifying agent. Examples of parenteral emulsions include:

- Oil in water emulsion (for sustained-release depot preparation given IM).
- Water in oil emulsion (allergen extract).
- Chylomicra: Lipid emulsion is the most widely used class of emulsion. Chylomicra are 0.5 to 1.0 μm spheres consisting of central core of triglycerides and the outer layer is made up of phospholipid. IV emulsion usually contains fats up to 10%; this ratio can be increased upto to 20%. The emulsions containing fats can be used as TPN for providing nutrients and calories [4].

11.2.3 Formulation Considerations for Parenterals

While designing formulation, we need to consider different factors. Excipients are one of the most important factors among them. General classes of excipients used in parenteral formulation are summarized in Table 11.2. Certain excipients are unique to each category of parenteral formulations. However, there is a general understanding for the main factors that need to be considered while designing formulation for parenteral products [5].

Table 11.2 Common excipients used in parenterals

Excipients	Examples
Vehicle	*Aqueous-based vehicle*: e.g., water for injection USP, sterile water for injection USP, bacteriostatic water for injection USP, etc.
	Nonaqueous vehicle: e.g., corn oil, cotton oil, peanut oil, sesame oil, etc.
	Co-solvents: e.g., glycerol, propylene glycol, ethanol, etc.
Buffers	Sodium citrate/citric acid and sodium acetate/acetic acid
Surfactants	*Nonionic surfactants (only)*: Tween®series, poloxamers, etc.
Preservatives	In multidose containers and non-terminally sterilized products, methyl and propyl para-hydroxybenzoic acid in low concentration of about 0.2% w/v, etc.
Antioxidants	Sodium bisulfite, sodium metabisulfite, ascorbyl palmitate, ascorbic acid, etc.
Tonicity modifiers	Sodium chloride, dextrose, mannitol, etc.

11.2.3.1 Type of Parenteral

Various types of preparations like solutions, suspensions, emulsions, etc. are there. Depending on type of preparation, we choose the suitable excipients. Selection of excipients depends upon various factors, i.e., the solubility of therapeutic agents, the route through which parenteral is being administered, amount of volume, and onset of action [5].

Solubility of Drug

Solubility is important because if drug is soluble in a vehicle or a solvent, it is easy to prepare a solution. If a drug is insoluble, we will consider suspension. Sometimes co-solvents are used in case when drug is insoluble in water, while considering the co-solvents, we need to focus on their toxicity because some of the co-solvents like ethanol are toxic if used in a higher amount [5]. Sometimes it happens that the drug is dissolved in a co-solvent system, but after some time the drug recrystallizes to its solid form. So, it should be confirmed that the drug will remain soluble in the used co-solvent system throughout its shelf life [5, 6].

Desired Routes of Administration

For IV route, mostly solutions are prepared using different excipients, while for subcutaneous or intramuscular route, we select the emulsions. But for emulsions, the size of droplet is kept below 1 μm because bigger droplet can irritate blood vessels same as the case for suspensions [5].

Volume Be Administered

Large volume parenterals up to 500 mL are administered intravenously, while the small volume parenterals can be administered through any route depending on type of preparation (either it is solution, suspension, or emulsion) [5].

Onset of Action

The most rapid action is produced by IV, compared to SC and IM. Aqueous solution has rapid onset as compared to oil-based solution or oil-based suspension or aqueous suspension. As for a suspension the drug must be dissolved first into solution form and then it will be active for pharmacological action [5].

11.2.3.2 Properties of Drug

For development of stable and safe parenteral products, the properties of drug need to be thoroughly investigated [5]. The physical and chemical properties of drug may affect the choice of formulation components for parenteral products. The important factor that affects the decision-making in formulation development process includes crystal characteristics/polymorphism, solubility, ionization constant (pKa), and particle size [7].

11.2.3.3 Properties of Vehicle

Vehicle is a medium which provides a medium to a system to carry a drug. There are three different types of vehicles: aqueous-based vehicles, oil-based vehicles, and hydroalcoholic vehicles [7].

Aqueous-Based Vehicle

Aqueous vehicles are used for freely soluble therapeutic agents like the drug which is highly soluble in water. Such drugs can be easily dissolved and form solutions in water [5]. For the drug with low aqueous solubility, co-solvents are used. Water is also used as an external phase for emulsions in which oil droplet is suspended in water by using various emulsifying agents. In USP, there are three different types of water mentioned used in pharmaceutical preparations, i.e., water for injection, sterile water for injection, and bacteriostatic water for injection [8].

Water for Injection USP Water for injection is non-sterile and used for large volume parenterals. After the product formation, these are terminally sterilized (i.e., when all the production processes are done and packed in plastic pouch-type container) by using autoclave or radiations. According to different pharmacopeias, water for injection has various limits regarding purity, appearance, amount of dissolved solids, pyrogens, and sterility. Pyrogens are substances that cause fever in a body. The most common pyrogens are endotoxins, which are lipopolysaccharides produced by microorganisms [7]. Pyrogens can be removed by heating at 250 °C for 30–45 min mostly or 180 °C for 3–4 h using autoclave. The most common method used to remove pyrogens from large volume parenterals is distillation or reverse osmosis. In the distillation process, water is simply boiled, all the solids and contaminants are left behind, and the steam moves to condenser where it converts back to pure water and used for parenteral preparations. In reverse osmosis, a semipermeable membrane is used which can segregate all the contaminates and dissolve solids in water to give pure water [8].

Sterile Water for Injection USP Sterile water for injection is sterile and used for reconstitution of drugs in powder form, e.g., ceftriaxone is in powder form packed along with an ampule of water on which it is mentioned that "sterile water for injection" is used for the reconstitution of ceftriaxone to administer parenterally [8].

Bacteriostatic Water for Injection USP This contains antimicrobial agents and most commonly benzyl alcohol in 0.9% w/v proportion. It is mostly used in multi-dose containers where the chances of microbial growth increase due to instant pricking.

Nonaqueous Vehicle

Nonaqueous vehicle is used where the drug is insoluble in water so oil is used for the drug dissolution. Nonaqueous vehicle can be either solution or suspension or as a droplet dispersed in an aqueous medium. Most common nonaqueous vehicles are corn oil, cotton oil, peanut oil, and sesame oil (these oils are very stable against oxidation so less or no chances of rancidity). Mineral oils cannot be used in parenteral solutions. Two problems occur with oily solution or suspension because of their high viscosity, i.e., irritation to the muscles where administered and pain, and in some patients there is a chance of sensitivity (some patients are sensitive to certain oils). Sometimes therapeutic agents are actually used as vehicle, e.g., benzyl benzoate is used as a nonaqueous vehicle and itself it is a therapeutically active compound [9].

Co-solvents

To increase the solubility of drug sometimes co-solvents are used. The common co-solvents include glycerol, propylene glycol, and ethanol. Co-solvents should be used in such concentrations that can assure the solubility of a product throughout the shelf life [5].

11.2.3.4 Surfactants

Surface-active agents are amphiphilic compounds used for solubility enhancement or to stabilize the particles used in suspension. Surfactants are used in concentration above the CMC (critical micelle concentration) to increase the solubility. Amphiphilic polymers (surfactants) form micelle when meet with physiological medium. If concentration is higher than CMC, it forms micelles, but if concentration is lower than CMC, it does not form the micelle rather deposits on the surface of particles which are suspended in the suspensions [5]. Nonionic surfactants are more commonly used in parenteral formulations. Examples include polyoxyethylene sorbitan fatty acid esters like Tween series (Tween 20, Tween 40, etc.) in concentration of 0.1–0.5% poly(oxyethylene)-poly(oxypropylene) like poloxamers in concentration of 0.01–5% [5].

The type of surfactant used in formulation depends on the hydrophilicity and lipophilicity of the surfactant. The surfactants are amphiphilic compounds having both hydro- and lipophilic nature. Each surfactant has some value of hydrophilic-lipophilic balance, known as HLB. If HLB value is more than 10, this denotes surface-active agent is hydrophilic in nature, and these will be used for hydrophilic vehicles or hydrophilic mediums, while if HLB value is less than 10, it means the surface-active agents will be used for lipophilic medium. For solubilization, higher concentration, i.e., more than CMC, is required [10].

Sometimes combinations of surfactants are used, and these types of delivery systems are known as self-emulsifying drug delivery systems (SEDDs). Examples include amphotericin B as a complex with sodium deoxycholate (Fungizone) and sodium cholesteryl sulfate (Amphocil). When the powder is constituted with water, it forms a colloidal solution before intravenous infusion [11].

11.2.3.5 Buffers

Buffers are combination of a weak acid and its salt or a weak base and its salt. Buffers are required because pH is required to be maintained, not only for biocompatibility but also for the stability of the drugs. Some of the drugs might degrade at a certain pH; so the pH should be maintained such that it is biocompatible and to ensure stability throughout shelf-life [5]. Thus, buffers enhance the chemical stability. Examples of buffers used for parenterals include acetic acid/sodium acetate and citric acid/sodium acetate [9].

11.2.3.6 Preservatives

Preservatives are most commonly used in multiple dose preparations, since each time a dose is taken out, therefore chances of contamination are there. The product is not terminally sterilized. Examples are esters of para-hydroxybenzoic acid like methyl and propyl para-hydroxybenzoic acid in low concentration of about 0.2% w/v [5].

11.2.3.7 Osmolarity Adjustment

Osmosis is the movement of water through a semipermeable membrane. Movement of water takes place as long as concentration gradient is present. When the equilibrium is achieved, the molecules of water at both sides are the same and no further movement takes place. This movement can be stopped by applying pressure; the minimum pressure required to stop the movement across the membrane is termed as osmotic pressure. If we have high concentration of solute, then we need high pressure to stop the movement of water which means the osmotic pressure of solution is higher. If the concentration of solute is low, less amount of pressure is required to stop the movement of water; thus the solution will have low osmotic pressure. This all depends upon the concentration of solute in a solution [5].

A solution will be considered isotonic if it has the same osmotic pressure to that of the blood. So ideally the solution administered parenterally should be isotonic. If the solution is hypertonic, then there will be movement of water from the blood to the product, and if the solution is hypotonic, then there will be movement of water from the product to the blood. Therefore, it is important that the solution to be administered must be isotonic [2] (Table 11.2).

11.2.4   Manufacturing of Parenterals

11.2.4.1   Manufacturing of Injectable Solution

By using a suitable vehicle drug, excipients are dissolved. Water for injection is the most commonly used vehicle for preparation of injectable solutions. Injectable products are most common and can be administered by any route [5].

Injectable solutions are manufactured by dissolving the drug and excipients. Then the pH of solution is adjusted because after addition of some excipients the pH of the solution can be changed. To maintain a constant pH, some buffers are added. The isotonic solution is now filtered through 0.22 μm membranes. These membranes have small pores. The sterile solution is then filled aseptically for thermolabile substances and autoclaved for thermostable materials. For multiuse containers, antimicrobials are added. The total filled volumes of fluid into a parenteral container are greater than labeled volume [12].

11.2.4.2   Manufacturing of Injectable Suspension

Suspensions are the products in which drug molecules are dispersed as a small particles (the drug does not dissolve as in solution form).

Aqueous vehicle is prepared containing all the excipients. Then active ingredients are suspended as a particle. Two methods are used for the preparation of injectable suspensions [12].

Incorporation of Sterile Powder in Sterile Vehicle

This is the most widely used method for preparation of injectable suspensions. In this method sterile powders are mixed and dispersed in sterile vehicle. Examples include Penicillin G Procaine injectable suspension USP sterile vehicle and all the soluble excipients (including lecithin, sodium citrate, povidone). All these are dissolved in vehicle and passed through 0.22 μm membrane. The solution is then transferred to pre-sterilized mixing filling tanks, and then the powder which is made sterile by freeze-drying and spray-drying, is added into already prepared vehicle in a sterile manner [5].

Sterile In Situ Crystallization and Mixing with Sterile Vehicle

This is another method for preparation of injectable suspensions. In this method sterile in situ crystals are taken and then these crystals are mixed with sterile vehicle. Example is sterile testosterone injectable suspension USP. First, we prepare a vehicle which is sterile and filtered through 0.22 μm membrane. Testosterone is dissolved in acetone separately and then it is filtered through 0.22 μm membrane. The testosterone sterile solution is added to sterile vehicle (already prepared). The drug is insoluble in the vehicle, so it crystallizes. The suspension is brought to desired volume and filled in normal prescribed manner [13].

For injectable suspensions, flow property is very important factor. Like if we observe the buffer oral suspension, it is quite viscous and such type of consistency is not acceptable for injectables. The flow property is usually characterized in terms of syringe ability and injectability. Syringe ability and injectability is defined in terms of withdrawal from container into syringe, and there will be no clogging and foaming and accurate dose measurement will be possible. Like if we have a container of 5 ml, once we draw the dose, it comes out 4.5 ml or 4.8 ml and it causes dose variations [13].

11.2.4.3 Manufacturing of Injectable Emulsions

Two types of emulsions are commonly prepared: oil in water and water in oil emulsion. Example of W/O emulsion is preparation of allergenic extracts like vaccines. In W/O emulsion external medium is oil while the droplets which are dispersed are water [13]. This emulsion is administered subcutaneously; depot preparations are administered in the muscles in bolus where it releases slowly to surroundings and then distributed throughout the body. Emulsifiers are used less commonly because of problem encountered like autoclaving (some of the emulsifier cannot be autoclaved), and some of the emulsifiers are toxic and can produce unwanted physiological effects; hence these properties limit the use of emulsifier [14].

The most common emulsion administered intravenously is chylomicron which is lipid emulsion and in use for the last 25 years. It has a very fine droplet size of 0.5–1 μm, the central core of droplet is made of triglyceride, and outer core is of phospholipid. It can administer about 20% of fats. These kinds of emulsions are usually used for patients who require nutrients in sufficient amount and cannot take it orally. It provides a lot of fatty acid and calories during TPN [14].

11.2.5 Sterilization of Parenteral Formulations

Sterilization may be defined as "the absence of living microorganisms (either through the destruction of all living microorganisms or by their removal) in pharmaceutical preparations" [15]. Five different methods are used for sterilization of pharmaceutical raw materials and pharmaceutical preparations. Different methods of sterilization are summarized in Table 11.3.

Table 11.3 Different methods of sterilization

Moist heat sterilization	
Procedure	In this method, pharmaceutical products are exposed to specific temperature for specific time, which results in efficient sterilization. For example, at 103.4 kPa pressure (i.e., 15 pounds per square inch) and 121 °C temperature, sterilization is achieved in 20 min, whereas at 68.91 kPa pressure (10 pounds per square inch), sterilization is achieved in 30 min.
Mechanism	Denaturation/coagulation of microbial proteins.
Applications	This method is used for sterilization of such materials which are both thermostable (within the conditions of the sterilization cycle) and through which moisture can perfuse.

Dry heat sterilization	
Procedure	It is performed at higher temperature and longer time as compared to moist heat sterilization, e.g., exposure at 170 °C for 1 h or 160 °C for 2 h or 140 °C for 4 h.
Mechanism	Exposure to higher temp for longer time, microorganisms are destroyed due to cellular dehydration and then pyrolysis/oxidation.
Applications	This method is employed for sterilization of such thermostable materials/products that cannot be readily sterilized by moist heat, e.g., heat-stable drugs/excipients, nonaqueous vehicles, such as oils, glycerin, propylene glycol, and glassware (e.g., bottles).

Filtration sterilization	
Procedure	This method involves the removal of microorganisms from solutions through sterilizing filters having pore diameter of 0.22 μm.
Mechanism	Very small pore diameter (0.22 μm) is sufficient for entrapment of bacteria and fungi. After using the filters (which contain the entrapped/retained microorganisms), these are then safely discarded.
Applications	This method is used for sterilization of thermolabile therapeutic solutions.

Radiation sterilization	
Procedure	In this method raw materials/products are exposed to a defined dose of ionizing radiation. Usually, gamma radiations in the range of 25–40 k Gy are employed for sterilization.
Mechanism	Microorganisms are destroyed due to exposure to gamma radiations.
Applications	This method is used for sterilization of therapeutic agents/excipients or parenteral formulations that are manufactured and packaged under aseptic conditions but are neither terminally sterilized nor sterilized by filtration.

Gas sterilization	
Procedure	In this method, materials/products are exposed to mixtures of ethylene oxide or propylene oxide along with an inert gas, e.g., carbon dioxide. This is done within a specially designed apparatus.
	Sterilization efficiency may be enhanced in the presence of moisture (up to 60%) and elevated temperature (55 °C).
Mechanism	The reactive gases destroy microorganisms by chemical reaction with cell proteins and DNA.
Applications	It is used for sterilization of medical devices and surgical accessories, e.g., packaged catheters and blankets. Therapeutic agents/excipients may also be sterilized by this technique.

11.3 Sterile Ophthalmic Preparations

Development of therapeutics for treatment of ocular disease is a challenging task, because ocular tissue is one of the most sensitive tissues of the human body. Moreover, many challenges need to be circumvented for improved bioavailability of drug through ocular route. For instance, frequent blinking, rapid tear turnover, and nasolacrimal drainage are some of the challenges that limit the bioavailability of topically applied ophthalmic products.

It is important to understand the anatomy of the eye, in order to know where the drug is applied in ocular tissue [16] (Fig. 11.1).

11.3.1 Anatomy of the Eye

Eye is a ball suspended in the ocular orbit and is composed of multiple tissues that coordinate to focus, transmit, and detect incoming light. The central path is transparent, the aqueous humor. The light travels through it and reaches the retina, that with the help of photoreceptors is converted into image [17].

Cornea: The outermost layer of the eye is cornea, which is a transparent membrane on the surface of the eye. This is the first and most important barrier of the eye. Cornea consists of three layers: the epithelium, stroma, and endothelium. These layers vary in their barrier properties. Cornea has no blood vessels so the nutrients are taken from the surrounding aqueous humor.

Sclera: Another tough part of the eye is sclera which is like a fibroblastic capsule, which encloses the eye and provides support as well as protection to the interior structure. Sclera is connected to the cornea through limbus [16].

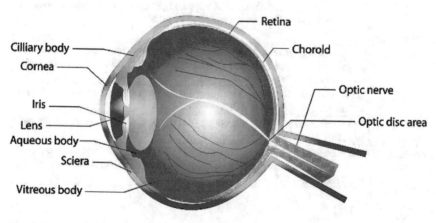

Fig. 11.1 Anatomy of the eye

The limbus: It is rich in blood vessels. Most of the absorption from topically applied ophthalmic products occur from this part of the eye.

The interior side of limbal area has a small opening, called canal of Schlemm, which helps to regulate intraocular pressure [17].

Iris: Next, there is *iris*, which is a ring of muscular tissue present in front of the lens, and it segregates the anterior and posterior chamber. It controls the entry of light into the back of the eye through the pupil.

Lens: It is a flattened sphere that is connected to the ciliary body by a fiber-like strand.

Vitreous humor: It is a transparent gelatinous material and has no turnover. It is in direct contact with the retina.

Retina: It is bilayer and a highly metabolically active tissue that is responsible for transforming light into an electrical signal that is converted to image by the brain [17].

11.3.2   Topical Ophthalmic Products

Topical ophthalmic products are applied to the surface of eye, the cornea. The common topical ophthalmic products are eye drops, ointments, and gels. They are typically used for treatment of diseases of anterior segments, for example, the different corneal layers, conjunctiva, sclera, iris, ciliary body, etc. [17]

11.3.2.1   Formulation Considerations

The bioavailability from topical products is affected by many factors, such as continued turnover of tears, nasolacrimal drainage, etc. The absorption from corneal surface can be increased by:

– Increasing the residence time of drug on the corneal surface
– By enhancing the permission of the drug through membrane

The residence time of drug in the eye can be prolonged by increasing the viscosity. However, viscosity higher than 70cps can cause blurring of vision and discomfort [17].

Drug permeation through ocular barrier can be increased by using lipophilic prodrug. For example, epinephrine has limited permeation through the membrane; however, its lipophilic prodrug, dipivefrin, can easily pass the corneal epithelium that is hydrolyzed to the parent epinephrine, after crossing the membrane. Moreover, permeability can also be increased by the use of penetration enhancers.

The osmolarity of the product is also a critical parameter to be maintained. Ideally, ophthalmic product should be isotonic, i.e., 300 milli osmols/liter). Very high hypertonic solutions can cause discomfort and induce irritation and lacrimation; hence the drug will be drained off.

The optimum pH value for ophthalmic products is 6–8. Very acidic and basic products can cause irritation and lacrimation. The increase in tear production will drain the drug off the corneal surface [18].

11.3.2.2 Classes of Topical Ophthalmic Product

The common classes of topical ophthalmic products are eye drops, eye ointment, gels, and gel-forming solutions.

Eye Drops

One of the most common topical ophthalmic products is eye drops. It is applied to the cul-de-sac. Usually more than one drop is instilled into the eye. The commercial ophthalmic droppers usually deliver 30–50 µL of fluid depending on fluid viscosity. It should be noted that in normal conditions eye can accommodate 10–15 µL of fluid. Typically, two to three drops are administered; the excess amount is drained off from the eye. Therefore, frequent administration is required. Eye drops can be in the form of solutions, suspensions, or emulsions [18].

Solution

Solutions provide uniformity of dose, improved bioavailability, and easy production process. Most of the drugs intended for ophthalmic administration are water soluble; therefore, it is easier to formulate in the form of solution. However, in case of water-insoluble drugs, salt formation is a common approach. The most common salts are hydrochloride, sulfate, nitrate, and phosphate. It is important to note that salt formation may change the physicochemical properties of drugs. For instance, different salts of the same drug pose different levels of irritation, as shown in Table 11.4.

Gel-Forming Solutions

These solutions tend to form gel after coming in contact with eye. Gel formation increases the residence time of drug on the absorption surface which in turn increases absorption. For example, timolol ophthalmic solution is applied twice a day for management of glaucoma, while the gel-forming solutions require only once a day administration. There are different mechanisms of gelation [19].

Table 11.4 Buffer capacity and ocular discomfort with different salts of epinephrine

Salt	Ocular discomfort level	Buffer capacity
Epinephrine bitartrate	Moderate to severe stinging	High
Epinephrine hydrochloride	Mild to moderate stinging	Medium
Epinephrine borate (decreased stability)	Occasional mild stinging	Low

Thermosensitive gelation: The product is solution at ambient conditions and forms gel at ocular temperature.

pH-sensitive gelation: Some solutions undergo gelation at ocular pH (i.e., 7.0 to 7.3).

Ion-sensitive gelation: Ocular fluid has a certain ionic strength. Some of the polymers form gels in the presence of certain ions or enzymes, such as lysozymes. For example, Timoptic-XE® uses gellan gum, which forms gel in response to higher ionic strength of tear fluid (US patent 4861760).

Another product uses xanthan gum, which forms gel in response to lysozyme (US patent 6174524) [19].

Suspensions

Suspension is recommended when the drugs are not soluble. The drug particle is dispersed in suitable vehicle. The particle size in ophthalmic suspensions is typically less than 10 μm, since bigger particle can produce irritation. Ophthalmic suspension should have good flowability and be easily resuspended upon shaking [19].

Ophthalmic suspensions are also used to increase the compatibility of the drug. Some drugs might have irritating property in solution form but nonirritating when administered as suspension. For example, betaxolol (beta-blocker) used in glaucoma produced a serious discomfort. Many solution-based formulations of betaxolol (β-blocker) were tried but showed limited success. Insoluble form of betaxolol (Betoptic®S), using 20 μm particles of high molecular weight polyanionic polymer carbomer and a sulfonic acid cation exchange resin, improved efficacy and ocular tolerance (US patent 4911920) [20].

Powders for Reconstitution

The drug that is unstable in solution or suspension form is formulated as powder that needs to be reconstituted before administration. These powders are usually manufactured by lyophilization (freeze-drying). The solutions are added into individual glass vials that are freeze-dried to form dry powder. Lyophilization requires bulking or stabilizing agents like mannitol and potassium acetate. A separate sterile vehicle should be dispensed along with powder for reconstitution [18].

Ocular Inserts

Ocular insert is a sterile preparation covered in a thin multilayered solid or semi-solid device designed to be placed in cul-de-sac to be in contact with the eye for longer period to provide a constant bioavailability. It prolongs the contact with corneal surface; hence the therapeutic activity of drug can be improved. To achieve this viscosity, enhancing agents are added to eye drops to increase the retention time of drug at eye surface, but these dosage forms only give sustained drug eye contact for specific period and do not yield a constant bioavailability [21]. Apart from that, several other approaches have been adopted for increasing corneal residence time of drug, such as microparticles, gels, gel-forming solutions, micelles, liposomes, nanoemulsions, NPs, ocular inserts, etc. [22]

Among these approaches ocular inserts have gained special interest for prolonging ocular residence time of drugs and reliably providing controlled release without any irritation to patient [23].

Soluble Ocular Inserts These are also called erodible ocular inserts. Polymers are being added in ophthalmic solutions to increase the viscosity and retention time. These are usually made of cellulose and water-soluble polymers, which can be sterilized by gamma radiations [23]. These are inserted into cul-de-sac of the eye for the improvement of absorption, enhanced bioavailability, and therapeutic effect. Since these are made with water-soluble polymers, so with the passage of time, these slowly dissolve and erode releasing their drug content. The potential advantage of erodible ocular inserts is not to remove at the end of useful dosing interval [23, 24].

Insoluble Ocular Inserts These are also called non-erodible ocular inserts. The first controlled topical dosage form was designed in 1975 by Alza Corporation. The insoluble inserts have been divided into three different groups. These include diffusion systems, osmotic systems, and hydrophilic contact lenses. The first two classes contain a reservoir, which is in contact with the inner surface of the retina. The reservoir contains a liquid, a gel, a colloid, a semisolid, a solid matrix, or a carrier-containing drug homogeneously or heterogeneously dispersed or dissolved therein [25].

11.4 Environmental Control for Sterile Manufacturing

Manufacturing of sterile pharmaceuticals and medical devices requires *clean room*, which is a controlled environment where the limit for the number of specified size is properly maintained [26]. The construction of the rooms for sterile manufacturing needs a proper design which minimizes generation, introduction, and retention of the particles [27]. Furthermore, other key parameters which include temperature, humidity, air flow filtration/velocity, and pressure are the trigger factors to be controlled for ensuring production of the aseptic products [28].

For parenteral drug manufacturing, environmental control is the major challenge to be addressed, because quality of the final products is directly influenced by the processing environment. The objective of environmental control in a sterile facility is to minimize the presence of all contaminants (both viable and nonviable) [29].

Microorganisms and dust particles suspended in the air are most likely to gain access to the product; therefore, an adequate environmental control program is needed. The following are the most imperative facilities that are required for ensuring and controlling the aseptic environment for the sterile products [30]:

1. Heating, ventilation, and air conditioning (HVAC) system
2. Personnel contamination control systems
3. Cleaning and disinfection of the area
4. Environmental monitoring systems

11.4.1 Heating, Ventilation, and Air Conditioning (HVAC) System

HVAC systems are considered the most imperative component of the environmental control systems design [31]. The main objective of HVAC system is to ensure the required standard environmental conditions for manufacturing of the pharmaceutical products [32].

Several components of an HVAC system can be segregated as follows:

- Blowers and fans for air generation.
- Cooling, heating, humidifying, and dehumidifying system for air conditioning.
- Metal ducts for air distribution networks.
- Air filtration equipments, depending upon the type of manufacturing. For sterile manufacturing, HEPA filters are installed.

For designing an effective HVAC system, some of the operational parameters need to be defined, since these parameters are considered critical for aseptic processing.

11.4.1.1 Temperature and Humidity Control

Temperature control is required to offer a comfortable working environment for operator. Usually 19–23 °C is considered acceptable [26].

Within manufacturing facility, some of the areas, like those where autoclaves, dry heat sterilization tunnels, and ovens are located, provide high heat loads on the systems. If not controlled, these loads can result in an environment which may not only cause discomfort for the working personnel but also increase the contamination chances due to high perspiration of operators [33].

Humidity control is also being considered the key factor for quality manufacturing of the pharmaceutical products. However, this also depends on the product requirement. Comfort levels are in the range of RH 45–55%, whereas manufacturing process requirement can vary widely. Some products (freeze-dried products) are manufactured in controlled environment with a relative humidity range of only 15–30%. Normal humidity levels can be easily achieved with air conditioning systems. Humidity levels lower than the normal can be achieved by air dryers in the air supply system, which work on the adsorption principle.

11.4.1.2 Pressure Differential Control

Pressure gradient is a mean to prevent cross-contamination between environments. Pressure gradients are established so as to provide critical environments with higher pressures than those that are less critical for the process. To increase the pressure of any environment, sufficient air must be added to ensure an overflow to the adjacent environment.

Differential pressure among various rooms or environments will be relative, i.e., one room will be measured relative to another.

11.4.1.3 Airborne Contamination Control

The HVAC system should provide control over the airborne viable and nonviable particles so as to satisfy the level of control specified for the environment. To prevent airborne contaminants from entering the clean or aseptic environment, all air supplied for the environment must be filtered [34].

One of the primary functions of the filtration system is to provide sufficient level of cleanliness by recirculating the air contained within the environment through various filters, mainly HEPA filters, thus providing a polishing effect [35]. This capacity is commonly defined by the number of air *changes* per hour or by the recirculation ratio of the system.

Air Changes Rate (ACR)

The frequency of air turnover in the clean room. It is the measure of how quickly the air in an interior space is replaced by fresh air:

$$Air\,change\,rate = \frac{Volume\,of\,airflow\,per\,hour}{Volume\,of\,room}$$

$$(11.1)$$

Air is distributed in such manner that it flows into the maximum security room at the greatest volume flow rate, hence producing positive pressure, which is successively reduced so that the air flows from the maximum security area to the other less critical areas for return to the filtration system.

The level and type of filtration needed depend on the level of cleanliness required. As a universal method for cleanliness classification, several industries have adopted a standard. This standard classifies cleanliness according to the number of particles, 0.5 micron or larger per m^3 of air as shown in Table 11.5.

Table 11.5 Limits for airborne particle in different classes of clean room [38]

| Grade | Maximum number of particles allowed per m³ equal or greater than the tabulated size | | | |
| | at rest state[a] | | in operation state[b] | |
	0.5 μm	5 μm	0.5 μm	5 μm
A	3520	20	3520	20
B	3520	29	352,000	2900
C	352,000	2900	3,520,000	29,000
D	3,520,000	29,000	Not defined	Not defined

[a]*at rest state* means the condition where the installation is complete and is operating in such manner agreed upon by the customer and supplier, but in the absence of any personnel
[b]*in operation state* means the condition where the installation is functioning in specific operating mode and in the presence of specific number of personnel

Fig. 11.2 Vertical laminar flow clean room

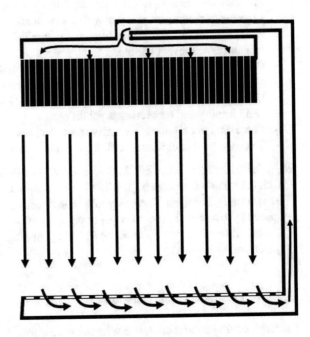

Laminar airflow (LAF) or unidirectional airflow has greatly improved the potential for environmental control of aseptic areas. Currently it is the only mean for achieving class 100 rooms. A laminar flow (unidirectional airflow) room is one in which air is introduced evenly from high efficiency particulate air (HEPA) (Fig. 11.2) filter panels and returned into the opposite surface of the room. The air moves in parallel flow lines with uniform velocity and minimum eddies. The air velocity across the room is held in the range of 90 ± 20 ft/min. Contamination is prevented because it is swept away with the airflow.

LAF rooms may have vertical flow, i.e., when the airstream is perpendicular to the floor, or horizontal flow, the air travels parallel to the floor. Vertical flow room offers a continuous clean air shower. Contamination generated in the area will be swept down and exhausted through a perforated floor. This type of room minimizes cross-contamination from one side of the room to the other in any direction. In a horizontal flow room, contamination generated downstream of the filter face will be exhausted across the room at the opposite wall. This type of room will minimize cross-contamination of operations taking place perpendicular to the airflow, but in operations taking place parallel to the flow, downstream contamination can occur, which is one of the limitations on the use of this type of room.

Advantages of LAF system: Unidirectional airflow possesses characteristics that make it one of the best contamination control systems. The following are some of the practical and technical advantages of LAF system:

- Particles less than 15 micron are swept at the same speed as the air; thereby particles of the largest group of airborne contaminants are removed.
- Air is filtered through HEPA-filtered air supplied, which provides the lowest contamination level possible.
- Particle removal efficiency of HEPA filter is 99.97% for particles 0.3 micron and larger. Thus, most bacteria are above 0.5 micron size, thus restrained.
- Due to the high recirculation ratio, unidirectional airflow systems provide the best cleaning and recovering capabilities.
- These are the only mean to achieve class 100 environments.
- Since the flow is unidirectional, the particles are not spread.

HEPA filter is a filter assembly that removes at least 99.97% of airborne particles with 0.3 micron in diameter [33]. The filter medium is composed of a folded mat of randomly arranged fibers typically composed of fiberglass. Every fold is spaced by grooved aluminum or paper separator. The particles in HEPA filter are trapped through a combination of straining, impingement, and entanglement (discussed in Sect. 4.3).

11.4.1.4   Personnel Contamination Control Systems

Excellent environmental control can be achieved only if the movements of personnel and supplies from one area to another are decreased [34]. It should be kept in mind that the access by personnel to the aseptic corridor and aseptic compounding and filling rooms is only through an *airlock*. For processing the aseptic products, the rooms with automatic lock door system are highly imperative and required. These will be very helpful to maintain the clean environment of the adjacent processing rooms. Furthermore, this system will prevent and control entrance of the microbes and particles from the non-sterile declared zones.

Personnel should be allowed to enter aseptic area only after following strictly the prescribed procedure for removing their street clothing, washing their hands, and wearing gowns, hats, shoes, face masks, gloves, and other prescribed outfits [29, 32]. After entering the area, they should not be allowed to move in and out of the area without re-gowning. Personnel working in this area should receive regular training on maintenance of discipline in clean rooms.

The lowest numbers of personnel required should be present in clean rooms. There should be minimum movement in the area, since a study showed that when we sit in working condition, human generate 0.5 million particles/min [36] of 0.3 μm and above. Similarly, one million particles/min and 5million particles/min are generated while standing and walking, respectively. Hence *penguin movement* is the suggestive walking style.

11.4.1.5   Cleaning and Disinfection of the Area

The cleaning equipment (e.g., mops) selected should be effective in cleaning and should not produce lint. The ceiling, walls, and other structural surfaces must be cleaned periodically [32].

All equipment and surrounding working area must be cleaned thoroughly at the end of the working day. No residue of the previous process should be present. After thorough cleaning, surfaces should be disinfected by either spraying or wiping and effective disinfectant on all surfaces. Irradiation from UV lamps is also sometimes used to reduce the viable microorganisms present in the area. UV lamps should be kept clean and checked periodically for its irradiation efficiency.

11.4.1.6   Environmental Monitoring Systems

The assessment of the level of environmental control can be performed by measuring total particle counts (viable and nonviable) in air sample, for example, the optical particle counter analyzer, which works on the principle of light scattering which is very useful to identify the number of particles in air sample [33, 37]. This device generates a plot, where the number of particles is displayed along with their respective particle sizes in the range of 0.3 μm up to 25 μm. The recommended particle size for the clean room, either equal to 0.5 μm or >0.5 μm, needs to be measured for quality compliance.

Viable counts are made using one or more of several methods, such as settling plates, contact plates, slit-to-agar samplers, or centrifugal samplers.

Settling plate sampling: In this approach, the uncovered petri dishes with the agar medium are placed on different spots for a specific period of time, where the microbe-bearing particles are more likely to be deposited on plate surface. It is a direct method of assessing the number of microorganisms depositing onto the product or surface in a given time.

Contact plate sampling: In this method, the agar plates are placed on the sampling spots for approximately 10 seconds in such a way that will interact with the surface evenly using a constant force. This will expose the maximum parts of the spots, where the microbes need to be collected from. Contact plate sampling is helpful for assessing the microorganisms on the surfaces of the building/machines where these organisms have either directly been deposited from the environment or from the contact by the workers and supervisors.

Slit-to-agar samplers: In this method, the air is introduced into the agar plates through a narrow slit, which whirls around the central axis and spreads across the entire surface. It is ideally required to cover the entire surface within 1 h, so the speed of the injected air will be fixed accordingly. This method is very handy to determine level of contamination at different periods of time during the ongoing activities.

Centrifugal samplers: This method utilizes an impeller device in the head of the air sampler to draw air. The impeller is then rotated to produce a centrifugal force which causes particles and microorganisms to impact onto an agar strip that is fixed around the border of the sampling head.

The results need incubation for 48 h at a specific temperature. In this time, the microorganisms multiply forming clear colonies. The number of microbes present in the sample is measured in terms of colony-forming units (CFU).

11.4.2   Classification of Clean Room

Sterile manufacturing involves different sections, such as compounding, filling, and packaging section. Each section requires different standards for environmental cleanliness.

Clean rooms are classified on the basis of cleanliness level in terms of viable particle and nonviable particles. Previously, the most universally applied classification was based on Federal Standard 209 of the US FDA. This standard was first published in 1963 in the USA. It was revised in 1966 (209A), 1973 (209B), 1987 (C), 1988 (D), and 1992 (E). However, FS 209E was officially cancelled by the US government on November 29, 2001 and was replaced with ISO 14644-1. The main limits are detailed in Tables 11.5 and 11.6.

Based on the required standard of cleanliness, clean room may be divided into critical area and supporting area. *Critical area* is the area for sterile manufacturing, where manufacturing containers, formulation ingredients, primary packaging material, and closures are exposed to environment. Critical area must be designed in such a way that it ensures sterility of the final products [39], while the *supporting area* is the area adjacent to the critical area.

For aseptically produced products, it is very imperative to control environmental cleanliness to the highest standard. However, *terminally sterilized products* require less stringent control compared to aseptic manufacturing (Table 11.7).

Aseptically prepared products: These are the products which are prepared within aseptic environment and are packed by closed system of **aseptic** transfer. It does not undergo sterilization step after packing.

Table 11.6 Limits for viable count in different classes of clean room [38]

| Grade | At rest state[a] | | In operation state[b] | |
	Air sample CFU/m³	Settle plates (diameter 90 mm) (CFU/4 h)	Contact plates (diameter 55 mm) (CFU/plate)	Glove print (5 fingers) (CFU/glove)
A	<1	<1	<1	<1
B	10	5	5	5
C	100	50	25	–
D	200	100	50	–

Table 11.7 Clean room requirement for aseptically prepared products and terminally sterilized products [40]

Sterile manufacturing	Clean room requirement	Pressure differential
Terminally sterilized		
Solution preparation	Class D	+15 pa next to outside
Component preparation	Class D	+15 pa next to outside
Filling (usually)	Class B[a]/C[b]	+15 pa adjacent area
Filling (risk of contamination)	Class A[a]/C[b]	+15 pa adjacent area
Aseptically prepared		
Solution preparation (needs filtration)	Class C	+15 pa next to outside
Product preparation	Class A[a]/B[b]	+15 pa adjacent area
Filling	Class A[a]/B[b]	+15 pa adjacent area

[a]Critical area
[b]Surrounding area

Terminally sterilized products: These are the products which are sterilized in the final container.

References

1. MacArthur RB, Rockwell K, Johnson A, Vaughan R, Coller BS. 4324 Phase 1 Sterile Product Formulation and Manufacturing at Academic Medical Centers: An Introduction for Translational Researchers. Journal of Clinical Translational Science. 2020;4(s1):42–3.
2. Broadhead J, Gibson M. Parenteral dosage forms. Pharmaceutical preformulation and formulation: CRC Press; 2016. p. 337–59.
3. Schaut RA, Weeks WP. Historical review of glasses used for parenteral packaging. PDA Journal of Pharmaceutical Science and Technology, 2017;71(4):279–96.
4. Akers MJ, Strother CS, Walden MR., (2014). Sterile Formulations. In: Fermentation and Biochemical Engineering Handbook (Third Edition); Elsevier, Amsterdam, Netherlands: p. 377-384.
5. Rayaprolu BM, Strawser JJ, Anyarambhatla GJDd, Excipients in parenteral formulations: selection considerations and effective utilization with small molecules and biologics. Drug Development and Industrial Pharmacy. 2018;44(10):1565–71.

6. van Hoogevest P, Liu X, Fahr A. Drug delivery strategies for poorly water-soluble drugs: the industrial perspective. Expert opinion on drug delivery. 2011;8(11):1481–1500.

7. Ain QU, Farooq MA, Caliskan B, Ahsan A, Aquib M, Hussain Z, et al. Stability Studies of Parenteral Products. Drug Stability and Chemical Kinetics: Springer; 2020. p. 247–63.

8. Gulati N, Gupta H. Parenteral drug delivery: a review. Recent patents on drug delivery Formulation. 2011;5(2):133–45.

9. Marschall C, Witt M, Hauptmeier B, Friess W, Biopharmaceutics. Powder suspensions in non-aqueous vehicles for delivery of therapeutic proteins. European Journal of Pharmaceutics. 2021;161:37–49.

10. Saengsorn K, Jimtaisong AJAPJoTB. Determination of hydrophilic–lipophilic balance value and emulsion properties of sacha inchi oil. Asian Pacific Journal of Tropical Biomedicine. 2017;7(12):1092–6.

11. Abdulkarim M, Sharma PK, Gumbleton MJAddr. Self-emulsifying drug delivery system: Mucus permeation and innovative quantification technologies.Advanced drug delivery reviews. 2019;142:62–74.

12. Muralidhara BK, Wong MJDdt. Critical considerations in the formulation development of parenteral biologic drugs.Drug Discovery Today. 2020;25(3):574–81.

13. Wang Y, Cai D, Chen C, Wang Z, Qin P, Tan TJBt. Efficient magnesium lactate production with in situ product removal by crystallization. Bioresource Technology.2015;198:658–63.

14. Goodarzi F, Zendehboudi SJTCJoCE. A comprehensive review on emulsions and emulsion stability in chemical and energy industries.The Canadian Journal of Chemical Engineering. 2019;97(1):281–309.

15. Vieregg JR, Martin SJ, Breeland AP, Weikart CM, Tirrell MVJPJPST. Inhibiting sterilization-induced oxidation of large molecule therapeutics packaged in plastic parenteral vials.PDA Journal of Pharmaceutical Science and Technology. 2018;72:35–43.

16. Alghamdi EAS, Al Qahtani AY, Sinjab MM, Alyahya KM. Extemporaneous Compounding of Ophthalmic Products. Extemporaneous Ophthalmic Preparations: Springer; 2020. p. 19–.19

17. Stjernschantz J, Astin M. Anatomy and physiology of the eye. Physiological aspects of ocular drug therapy. Biopharmaceutics of ocular drug delivery: CRC Press; 2019. p. 1–25.

18. Alghamdi EAS, Al Qahtani AY, Sinjab MM, Alyahya KM. (2020) Topical Ophthalmic Drug Forms. In: Alghamdi EAS, Al Qahtani AY, Sinjab MM, Alyahya KM., (Eds), Extemporaneous Ophthalmic Preparations: Springer Nature Switzerland: p. 17-50.

19. Reed K, Berger NJIJPS, Res. The effect of polyvinylpyrrolidone (PVP) on ocular gel forming solutions composed of gellan and calcium gluconate. Int J Pharm Sci and Res.2018;9:20–8.

20. Müllertz O, Jacobsen J, Thyssen JP, Horwitz A, Kolko MJRMC, Images, Videos. Betaxolol Ophthalmic Solution as Alternative Treatment for Patients with Timolol Allergy: A Case Report. Reports.2020;3(3):21.

21. Gajanan GV, Pravin P, Adhikrao YJAJoP, Technology. Methods for Evaluation of Ocular Insert with Classification and Uses in Various Eye Diseases-A Review. Asian Journal of Pharmacy and Technology. 2017;7(4):261–7.

22. Abbas MN, Khan SA, Sadozai SK, Wahab A, Khan F, Hussami S, et al. Gelatin-PAA Hybrid Nanoparticles for Sustained Release Drug Delivery against Conjunctivitis Causing Pathogen. Latin American Journal Of Pharmacy. 2020;39(12):2536–44.

23. Devhadrao N, Siddhaia M, Therapeutics. Review on Ocular Insert Drug Delivery System. Journal of Drug Delivery and theraputics. 2018;8(5-s):115–21.

24. Alvarez-Lorenzo C, Hiratani H, Gomez-Amoza JL, Martínez-Pacheco R, Souto C, Concheiro AJJops. Soft contact lenses capable of sustained delivery of timolol. Journal of pharmaceutical sciences. 2002;91(10):2182–92.

25. Di Colo G, Burgalassi S, Chetoni P, Fiaschi M, Zambito Y, Saettone MFJIjop. Gel-forming erodible inserts for ocular controlled delivery of ofloxacin.International journal of pharmaceutics. 2001;215(1–2):101–11.

26. Akers MJ. Sterile drug products: formulation, packaging, manufacturing and quality: CRC Press; 2016.

27. Wiker G. 11 Sterile Manufacturing Facilities. Good Design Practices for GMP Pharmaceutical Facilities.CRC Press. 2016:295.
28. Chapman DG., (2019). Production of sterile products. In: Watson J, Coga LS, (Eds), Pharmacy Practice 6th Edition, Elsevier, Amsterdam, Netherlands: p.182-194.
29. Ljungqvist B, Reinmüller B. Clean room design: minimizing contamination through proper design: CRC Press, Boca Raton, Florida, United States; 2018.
30. Alavi-Moghadam S, Sarvari M, Goodarzi P, Aghayan HR. The Importance of Cleanroom Facility in Manufacturing Biomedical Products. In: Arjmand B, Payab M, Goodarzi P. (Eds), Biomedical Product Development: Bench to Bedside: Springer Nature Switzerland; 2020. p. 69–79.
31. Patil AB, Karnalli AP. Validation and Various Qualifications in HVAC System-A Review from Pharmaceutical Quality Assurance Prospect. International Journal of Pharmaceutical Quality Assurance. 2018;9(03):281–6.
32. Sandle T. Establishing a Contamination Control Strategy for Aseptic Processing. American Pharmaceutical Review. 2017. https://www.americanpharmaceuticalreview.com/Featured-Articles/335458-Establishing-a-Contamination-Control-Strategy-for-Aseptic-Processing/ (Accessed 29 March 2022)
33. Hasan MM, Hossain MJ, Chowdhury S, Halim MI, Marjia U, Rashid HA. A novel proposition of facilities required for sterile pharmaceutical preparation. The Pharma Innovation Journal,2017.6(10), 125-146
34. Halls N. Effects and causes of contamination in sterile manufacturing. Microbiological contamination control in pharmaceutical clean rooms: CRC Press, Boca Raton, Florida, United States; 2016. p. 13–34.
35. Agalloco J, Akers J. The future of aseptic processing. Advanced Aseptic Processing Technology: CRC Press, Boca Raton, Florida, United States; 2016. p. 465–9.
36. Whyte W and Hejab M. Particle and microbial airborne dispersion from people. European Journal of Pharmaceutical Sciences 2007;12(2):39–46.
37. Yaman A. 2018. Engineering Considerations in Sterile Powder Processes. In: Avis KE (Ed) Sterile Pharmaceutical Products: Routledge; 2018. p. 269–304. https://doi.org/10.1201/9780203738320
38. WHO good manufacturing practices for sterile pharmaceutical products. In: WHO Technical Report Series; No 961, 2011. (Accessed 29 March 2022). https://www.gmp-compliance.org/files/guidemgr/WHO_TRS961_Annex06.pdf
39. Akers J, Izumi Y. Technological advancements in aseptic processing and the elimination of contamination risk. Advanced Aseptic Processing Technology: CRC Press, Boca Raton, Florida, United States; 2016. p. 416–22.
40. Boom FA, Le Brun PPH, Bühringer S. and Touw DJ. Microbiological monitoring during aseptic handling: Methods, limits and interpretation of results. European Journal of Pharmaceutical Sciences, 2020; 105540. https://doi.org/10.1016/j.ejps.2020.105540

Chapter 12
Hazards in Pharmaceutical Industry

Inzemam Khan and Syed Majid Shah

Abstract Pharmaceutical unit operations follow standard operating procedures (SOPs); however, there are still chances of multiple hazards that may occur either due to an accident or due to prolonged exposure to hazardous materials. Proper safety measures are needed to avoid any kind of hazard to all the personals and the property. This chapter presents the different industrial hazards that may damage to the product/property and the preventive measures against these hazards.

Keywords Noise hazard · Lighting hazard · Fire hazard · Electric hazard · Heat hazard · Mechanical hazard · Psychosocial hazard

12.1 Introduction

"Hazard is physical or chemical characteristics that adversely affect human health, propertyand environment." Human health hazards refer to the substances, incidents, circumstances and activities that may lead to injury, irritation, sensitivity, toxicity or other adverse health effects [1, 2].

I. Khan
University of Peshawar, Peshawar, Pakistan

S. M. Shah (✉)
Department of Pharmacy, Kohat University of Science and Technology,
Kohat, Khyber Pakhtunkhwa, Pakistan
e-mail: smajid@kust.edu.pk

12.2   Types of Industrial Hazards

The following are some of the most common health hazards encountered during pharmaceutical processing:

1. Physical hazards
2. Chemical hazards
3. Mechanical hazards
4. Psychosocial hazards

12.2.1   Physical Hazard

Physical hazards are environmental or naturally occurring processes which tend to harm workers without necessarily touching them. These hazards include noise, improper lighting, fire, electric, heat, radiation and pressure [2].

12.2.1.1   Noise

Noise is an unpleasant and disruptive environmental stressor. When the sound level exceeds 85 decibels in an industrial setting, then it is referred to as industrial noise [3]. According to the Occupational Safety and Health Administration (OSHA) guidelines, noise exposure of a worker averaged over 8 h must be less than 85 decibels and must not exceed 115 decibels at any time [4].

In the pharmaceutical industry, a high level of noise may be produced by manufacturing and packaging machinery. Personnel working in production area are more likely to be exposed to industrial noise [5].

Continuous exposure to medium- to high-level noise may lead to many deleterious health effects among which hearing loss is the most prevalent one [6]. Industrial noise is also associated with an increased risk of cardiovascular diseases, such as hypertension, tachycardia, ischemic heart disease and stroke. Noise can also have negative psychological effects, such as stress, insomnia, anxiety, irritability and mood swings [7]. Moreover, it is associated with masking of danger signals, i.e., audible alarms as well as increases response time so may lead to industrial accidents [8, 9].

Preventive Measures
- Preference should be given to machines that are low noise generating.
- All the machines and equipment should be operated and maintained according to manufacturer's advice.
- Vibration absorbent material, i.e., footing cushions and noise damping agents like paints, coatings, etc. and muffling devices should be considered for on-site engineering of machinery.

- High noise-generating machines should be enclosed, isolated, or placed in soundproof cabinets to limit the number of exposed workers.
- Personnel exposed to a sound level above 85 decibels should be equipped with protection accessories that must be comfortable to use in addition to serving the purpose, i.e., earplugs and ear muffles [10].
- Programmed job rotation of personnel working in noisy areas should be considered as it reduces exposure of workers to noise and consequently the risk of adverse effects may decrease.
- Audiometry should be performed every 6 months or annually for employees exposed to noise levels above 85 decibels and corrective measures must be taken if required.
- According to OSHA, comprehensive hearing conservation program should be implemented, which includes identification of noise source, reduction of noise level, and proper training of exposed workers to effectively protect themselves from ill effects of noise [9, 11].

12.2.1.2   Lighting

Improper lighting both under illumination and over-illumination in the industry contributes to the increasing frequency of accidents and health problems. Poor lighting makes it difficult to detect any danger (any small sharp object, etc.), increases falling incident and decreases proper judgment (i.e., position and speed of a machine), thus leading to an elevated risk of industrial accidents [12]. It is also associated with many medical symptoms, i.e., eye strain, headache and muscular fatigue. Since the worker must adopt uneasy positions to focus on the task, poor lighting may also lead to errors and poor performance, which may result in increased production cost [13].

Preventive Measures
- Proper illumination must be provided by using the most appropriate light source.
- Careful selection of the location of light source installation should be made to provide appropriate illumination without glare or shade.
- Appropriate combination of natural and artificial light should be provided.
- Light fixtures must be well designed and appropriately spaced for proper lighting and should be cleaned regularly.
- Blinds, curtains, shades and louvers may be used for windows or skylights to make the field of vision free from glare.
- Walls and ceilings should be painted with light colors and must be kept clean.
- Local lighting should be provided to workers to prevent eye strain and fatigue if needed.
- Eliminate flickering of florescent tubes by screening the ends of tubes from direct view [14].

12.2.1.3 Fire

Fire is the rapid oxidation of fuel which results in heat, light and flame. In an industrial setting, uncontrolled fire may result in direct and indirect loss. Direct loss includes human loss (i.e., injury or death) and property loss (i.e., burnt equipment, materials, damaged building). Indirect loss involves loss of orders, bad publicity due to fire incidents, administration costs and legal action against the industry [15]. Usually, this indirect cost is more troublesome and may lead to the shutdown of business. Therefore, fire safety management must be a serious concern to fulfill legal requirement, to ensure employees' safety and to assure the future of the industry [16].

In the pharmaceutical industry, certain processes such as extraction, uncontrolled chemical reactions and organic synthesis may lead to a fire. Secondary manufacturing operations (i.e., granulation, milling, drying, compounding) can also pose a risk of fire and explosion. Moreover, pharmaceutical dust produced during formulation processes is highly combustible and results in an explosion when the source of ignition is available. Other principal causes of industrial fire include overheated machinery and clogging of ventilation filters with combustible material [17]. Three types of fire are common in the industry. Class A fire is originated from solid materials (like wood, paper) and is controlled by water. Class B fire is originated by flammable liquids or gases (e.g., gasoline, fuel oil) and extinguishers for this type include carbon dioxide, dry chemical and halogenated agent. Class C fire is caused by energized electrical equipment and non-conductivity-inducing extinguisher is required for it [16].

Preventive Measures

- Careful designing of plant layout, selection of building material, proper ventilation, installation of sufficient fire alarms, provision of exit points of adequate capacity and availability of fire extinguishers and sprinkler system can limit the risk of fire and explosion hazard.
- Eliminate unnecessary ignition sources such as smoking, flames and sparks, etc.
- Inflammable products must be properly stored in a separate, adequately ventilated storage area or in suitable safety cabinets and the quantity in working area must be limited.
- All electrical equipment used in the industry must comply with specifications and be properly maintained.
- Use of friction-prone tools and spark close to combustible materials or ignitable materials in hot workplaces must be avoided [18].
- Properly clean and dispose of any spilled flammable liquids.
- Aisles, stairs, and exit points must be kept clear of any combustibles.
- Flame-resistant clothing should be provided to employees working in areas where the risk of flame or arc is high.
- Ensure proper housekeeping to diminish fire risk from airborne combustible dust.
- Establish fire hazard assessment, training of employees to work safely, fire emergency action plan and rescue and medical procedures [19].

12.2.1.4 Electric

Electric hazard is referred to a condition when injury, flash burn, thermal burn, or death occurs due to contact with an energized conductor or due to fire or explosion of electric origin. Although electrical hazard is not the leading cause of occupational injuries, it must be taken seriously because it may cause serious injuries, fatality, and property damage [20].

Electrical injuries can be classified into four types, i.e., electrocution (death or severe injury by electric shock), electric shock, burns, and falls caused due to contact with electricity. The severity of the injury is influenced by various factors such as current strength, duration of exposure, current type (AC/DC), frequency of supply, the current path through the body (near the heart or nerves, etc.), and the resistance of the body (depends upon age, gender, wetting of skin, insulated or non-insulated floor, etc. [21].

Sources of electric hazard in the industry include overloaded circuit, improper earthing, improper insulation, high voltage, inadequate or faulty wiring, contact with a live wire, insulation failure, overheating of current-carrying component, substandard equipment, improper maintenance, short circuit, overheating, or ignition of flammable materials placed close to electric equipment. Personnel working near electrical equipment and machines and maintenance staff are at a high risk of electric hazard [22].

Effects of electric shock include loss of muscle control, respiratory arrest, arrhythmias, conduction problem, myocardial damage, generalized weakness, extreme pain, autonomic dysfunction, memory problem, deafness, restlessness, paralysis, cataract, eye lesion, blindness, severe burns, dislocation of major joints, and fracture of long bones due to falls or violent muscle contraction [23].

Preventive Measures
- All machines must be properly earthed.
- Avoid the use of low-quality electric equipment.
- Avoid touching any electrical circuit unless it is switched off.
- Overloading of outlets must be avoided.
- Avoid placing cords near heat or water.
- Worn out and frayed cord must be replaced immediately.
- All the electrical equipment and wirings must be properly maintained.
- Explosion-proof devices and non-sparking switches should be used.
- Any potential electrical problem must be reported to the manager or supervisor.
- Proper earthing of high voltage equipment should be done before any maintenance work.
- Insulation of electric equipment and cables must be done with standard material [24].
- Protection devices such as fuse, circuit breaker, miniature circuit breaker (MCB), automatic voltage regulator and anti-electrocution devices must be used.

- If any equipment or machine is faulty under repair or maintenance, it must be locked out and tagged out.
- Personal insulation materials, i.e., rubber gloves, rubber shoes and suitable dry clothing and insulated tools should be used in areas with a possibility of electrical hazard [22, 25].

12.2.1.5 Heat

Exposure to excessive heat at work not only causes health and productivity issues but also contributes to workplace injuries [26]. While working in a hot environment, the body adopts some physiological mechanisms to maintain its internal temperature mainly through increased blood circulation to the skin and sweating. However, when the environment is too hot (close to or above body temperature) and humidity level is also high, then heat gain by the body becomes more than heat loss. This situation is called heat stress [27].

Heat stress may lead to heat-related disorders (such as dehydration, heat cramps, heatstroke, heat syncope) among which heatstroke is the most fatal one. Furthermore, the risk of injury is greatly increased as cutaneous vasodilation leads to reduced blood flow to the brain and muscles, resulting in decreased mental alertness, dizziness, inability to concentrate, fainting and fatigue [28]. Scalding and burns due to accidental contact with hot surfaces or steam may also occur. Besides bringing physiological changes, heat stress also affects employee's behavior. These changes include low morale, irritability, taking breaks, reducing activity, frequent errors, unsafe behavior, taking shortcuts and absenteeism to reduce heat exposure. Behavioral changes result in reduced productivity, safety, and individual economic outcome [26].

Preventive Measures
- Employer must ensure the provision of water, cool rest places for workers, suitable cooling system and ventilation at the workplace.
- Unscheduled breaks must be allowed to employees if they report feeling dizzy, weak, extremely exhausted, or confused.
- Workers should be acclimatized by gradually increasing their exposure to hot workplaces.
- Insulation and regular inspection of thermal fluid and steam pipelines must be done.
- Workers should be directed to stay away from pressure-release valves and steam vents. Hot workplace employees must be provided suitable insulated gloves, ice-cooled condition clothing, aprons or reflective clothing.
- In hot weather consider shortening of work hours, if possible; otherwise give short frequent breaks to shorten the duration of each exposure [29].
- Cooler part of the day should be selected for hot jobs if practical.

– Workers must drink plenty of water, learn symptoms of heat illness, wear breathable clothes, avoid direct heat when possible, self-monitor themselves for heat stress symptoms and report to the manager if they have any.
– Medical attention must be provided immediately if any employee shows symptoms of heat-related illness.
– Heavy tasks must be assisted using mechanical equipment.
– Minimize radiative heat gain by insulating, shielding, or decreasing emissivity of radiation sources and hot surfaces [26].

12.2.1.6 Radiation

In the pharmaceutical industry, personnel are at risk of overexposure to radiations as radiations are used for different applications, for instance, ultraviolet radiation is used for sterilization and synthesis of vitamin D from ergosterol [30].

Potential problems from radiation exposure include ocular injuries (i.e., eye strain, conjunctivitis, cataract, pterygium, photokeratitis, photo-retinitis, retinal burns, etc.), dermatological injuries (i.e., erythema, skin sores, skin cancer, skin aging), lip cancer, autoimmune diseases, and some viral disease activation [31]. It is reported that chronic users of drugs (e.g., phenothiazine, tetracycline) are more likely to be abnormally sensitive to UV radiation.

The severity of radiation injury depends upon the duration of exposure and dose and rate of radiations, tissue radio sensitivity, inflammatory response, comorbidities (diabetes, connective tissue disorders, etc.), age of the person, hereditary DNA repair defect disorders, intake of drugs (i.e. chemotherapeutic agents) which increases radiation sensitivity, etc. [32].

Preventive Measures
- Both NCRP and ICRP guidelines suggest justification, optimization, and dose limitation as basic principles for radiation hazard prevention.
– Minimize exposure and maximize distance from the source of radiation whenever possible.
– High-intensity radiation sources must be isolated, enclosed, or shielded.
– Radioactive material stored in a restricted area must be properly labeled, and entry of unauthorized personnel in such room or unauthorized removal of material must be prevented.
– Employers must identify and evaluate radiation hazards present in the industry and institute employee awareness and training programs to minimize risk.
– Employees at risk of overexposure to radiation must be provided with safety spectacles, protective gloves, lead apron, and radiation monitoring devices (i.e., pocket dosimeter, pocket chamber, film badges).
– Any incident of radiation overexposure must be reported within 24 h [33].

12.2.1.7 Pressure

A variety of pressure equipment are used in industry rendering industrial personnel at the risk of pressure hazard. The equipment includes steam boilers, pressurized hot water boilers, gas storage tanks, air compressors, chemical reaction vessels, autoclaves, cylinders, and pressure pipes or tubing. Most common hazards related to high-pressure vessels or systems include design or construction error, installation errors, vibration, leakage, vacuum, mechanical breakdowns, failure of safety systems, inadequate training of operators, human error (i.e., improper opening or closing of valves, overfilling of vessels), inadequate inspection and maintenance, and breakage of high-pressure pipes, hose, or tubing [34].

Rupturing of high-pressure gas pipelines may cause fire and explosion resulting in injuries to personnel in close vicinity and property damage. Sometimes this explosion may also result in a collapse of a building. Violent bursting of vessels due to over-pressurization inside the vessel occurring due to heat absorption by cryogenic liquid or failure of safety valves results in impact and fragment-related injuries. Dropping of cylinders also causes injury to nearby personnel along with damage to machinery and building [10]. Leakage of gases may render personnel at risk of suffocation, tissue and bone damage, chemical or thermal burns, and poisoning. Other harmful health effects of high pressure are dizziness, nausea, twitching and vision deterioration, confusion, convulsion and subsequent death.

Pressure hazards may be detected by employing numerous methods including sounds, leak detectors, scents, soap solutions and visual checks for contamination or corrosion [35].

Preventive Measures
- Proper training and supervision of workers to minimize operator error.
- Avoid keeping pressurized gas vessels near ignition sources such as boiler, furnaces, radiators etc.
- Safety valves should be used to avoid the over-pressurization of vessels.
- Pressure system and hoses should only be used under the manufacturer's recommended conditions.
- Pressurized containers should not be dropped or struck.
- Vibration dampening devices should be used to reduce vibration.
- Consider the reduction of joints in the pressure system to decrease the potential for leakage.
- Only adequately trained and tested employees should be allowed to repair, calibrate, or maintain a pressurized system.
- While handling, using or transporting pressure cylinder, safeguards should be used.
- Consider using shields or barricades around the pressure system and restricting the access.
- Ensure provision of safety goggles and face shield to personnel working with the high-pressure system [10].

12.2.2 Chemical Hazards

Exposure to hazardous chemicals is considered as one of the major occupational health and safety (OHS) problems in the pharmaceutical industry, since it can cause short-term or long-term adversities. Chemicals cause harm to the body either through direct contact with the skin or by entering through different routes, such as the nose, mouth, and eye [4].

Hazardous chemicals may be explosive chemicals (e.g., amyl acetate, phosphorus trichloride and ammonium nitrate) corrosive chemicals (e.g., amyl trichlorosilane, anisoyl chloride and bromine trifluoride), flammable chemicals (e.g., aluminum triethyl, sodium aluminum hydride and amyl chloride), poisonous chemicals (e.g., aniline, arsenate of lead, mercuric ammonium chloride and benzoate and mercuric cyanide), carcinogens (e.g., asbestos, acrolein, aniline and acrylonitrile), oxidizing agents (e.g., aluminum nitrate, silver nitrate, chlorate, chlorite and barium nitrate), and noxious gases (e.g., methane, butane and pentane) [36].

Preventive Measures
- Highly dangerous chemicals (e.g., carcinogens) that cause a serious threat to human health should be substituted with a less hazardous chemical.
- It must be ensured that the substituted chemical is devoid of dangerous health effects.
- Toxic chemicals should be kept isolated from other chemicals.
- These should be stored in well-organized control rooms or cabinets with the proper maintenance of premises.
- There should be the installation of proper buffer areas around such hazardous or toxic chemicals to avoid direct exposure to the environment.
- Such chemicals should be used with strict precautions and when people are not present in the vicinity [4, 37].
- The handling or transfer of flammable, explosive, volatile, or other hazardous liquids between the containers should be through closed sealed pipes rather than physical handling to avoid hazards.
- Local ventilation system, such as hood, duct and drain pipes, should be used to remove hazardous chemical fumes and dust.
- Proper knowledge and training should be provided to the workers or employees regarding the usage of the correct type of personal protective equipment and clothing for specific chemicals [1, 4] (Table 12.1).

12.2.3 Mechanical Hazard

The hazards caused by manual or energy-operated equipment, machines or tools are termed mechanical hazards. These mechanical hazards may lead to injuries and death. Injuries like crushed hands, fingers, burns, blindness, or amputation may occur to the operators or other workers in the area as a consequence of unsafe use of

Table 12.1 Globally harmonized system (GHS) for identification of chemical hazard

S. no.	Category	Symbol	Pictograms
1.	Flammable	Flame	
2.	Oxidizing	Flame over circle	
3.	Acute toxicity	Skull and crossbones	
4.	Explosive	Exploding bomb	
5.	Hazardous to the environment	Dead tree and fish	
6.	Health hazard/hazardous to the ozone layer	Exclamation mark	
7.	Serious health hazard	Health hazard	

machines or contact with sharp edges, moving parts, and hot surfaces of machines [38].

The following are some of the typical machine and equipment-related hazards:

- Part of the body (i.e., finger and hand) is entrapped between a moving part of a machine, belt pulleys, and a fixed structure.
- Cutting due to contact with cutting edge, i.e., cutting disc.
- Stabbing from sharp or pointed operating component protruding from the machine.
- Struck by a moving part of the machine.
- Part of machines or material under work may escape, fly off, or hit the body [39].

Causes of mechanical hazards include moving parts, cutting edges, protruding parts, falling objects, unexpected startup or movement of machinery, equipment overheating or catching fire, unsafe practice, faulty inspection, absence of lockout procedures, inadequate training, poor machinery design, machine failure, lack of supervision and rough and slippery surfaces [40].

Preventive Measures
- Implement lockout and tag-out procedures.
- Avoid contact with moving parts of machines or equipment and hot surfaces.

- Use of fragile glassware or damaged equipment must be avoided.
- Avoid wearing loose clothing and jewelry while operating machines.
- All the equipment and machines must be properly maintained and inspected regularly [41].
- Floor must be non-slippery and enough working area must be provided to employees.
- Provide enough space for employees.
- Machine guards (i.e., interlocked guards, fixed guards, adjustable guards), control switches, and emergency shutdown devices must be used to minimize hazard.
- Personal protective equipment, i.e., glove, should be used.
- Employer must provide training to employees regarding the proper use of machines and establish a plan to assess and manage mechanical hazards and to provide care to injured employees [42].

12.2.4 Psychosocial Hazards

Psychosocial hazards have the potential to adversely affect the mental health and physical well-being of workers. Psychosocial hazards may occur due to poor work design, disorganized working environment and financial along with social reasons [43]. Moreover, workplace violence is another common contributing factor to psychosocial hazards [44].

Violence may pose threat to victim's mental and physical health as they are more likely to suffer from mental illnesses, insomnia, gastrointestinal problems, blood pressure, cardiovascular events, depression, feeling of being helpless or insecure, and suicidal thoughts. Based on the mode of violence, it is categorized into four types, namely, physical violence, verbal violence, psychological violence and sexual harassment. Verbal violence is reported to be most common among these four types [45].

Preventive Measures
- Industries must develop and implement policies and procedures that ensure positive workplace culture, characterized by respect for every worker, recognition of good work, encouragement of employees, balanced power structure, conflict resolution, zero tolerance against violence and sexual harassment, supportive work environment, effective communication mechanisms and employee adequate training and skill development programs.
- At the time of hiring, criminal background and psychological screening must be performed.
- Employees should be allowed to be part of the decision-making process.
- Role of each worker must be clearly defined and unnecessary interference must be avoided.
- Surveillance cameras should be installed to keep an eye on employees' activities.

- Installing devices such as mirror glass, clear plastic partition, etc. in the work area should be considered as they may give employees complete view of their surroundings, thus rendering them at less risk of violence.
- Supervisors and managers must be trained to ensure a healthy work environment, resolve conflicts, and properly respond to violent behavior [46].
- Meaningful and effective sanctions must be imposed to discipline employees when they fail to comply with rules and regulations.
- Industry must offer an Employee Assistance Program (EAP), so employees may seek help.
- Any report against violence and harassment must be properly investigated and responded and appropriate and timely action must be taken against the convicted.
- Bringing pistols, guns, or any other weapon must be prohibited in the working environment [47].

References

1. Agarwal P, Goyal A, Vaishnav R. Chemical hazards in pharmaceutical industry: an overview. Asian Journal of Pharmaceutical and clinical research. 2018;2:28–35.
2. Hafeez A, Ahmad S, Al-Taie A, Siqqqui S, Talwar I, Kamboj A, et al. Industrial hazards and safety management in pharmaceutical industry. 2020.
3. Organization WH. Environmental noise guidelines for the European region. 2018.
4. Bhusnure O, Dongare R, Gholve S, Giram P. Chemical hazards and safety management in pharmaceutical industry. Journal of Pharmacy Research. 2018;12(3):357–69.
5. Raja RV, Rajasekaran V, Sriraman G. Non-auditory Effects of Noise Pollution on Health: A Perspective. Indian Journal of Otolaryngology and Head & Neck Surgery. 2019;71(2):1500–1.
6. Hughes P, Ferrett E. Introduction to Health and Safety at Work: for the NEBOSH National General Certificate in Occupational Health and Safety: Routledge; 2015.
7. Farooqi Z, Sabir M, Latif J, Aslam Z, Ahmad HR, Ahmad I, et al. Assessment of noise pollution and its effects on human health in industrial hub of Pakistan. Environmental Science and Pollution Research. 2020;27(3):2819–28.
8. Dehaghi BF, Khademian F, Angali KA. Non-auditory effects of industrial chronic noise exposure on workers; change in salivary cortisol pattern. Journal of Preventive Medicine and Hygiene. 2020;61(4):E650.
9. Basner M, Babisch W, Davis A, Brink M, Clark C, Janssen S, et al. Auditory and non-auditory effects of noise on health. The lancet. 2014;383(9925):1325–32.
10. Friend MA, Kohn JP. Fundamentals of occupational safety and health: Rowman & Littlefield; 2018.
11. Bell LH, Bell DH. Industrial noise control: Fundamentals and applications: CRC Press; 2017.
12. Hajibabaei M, Rasooli E. Comparison of different methods of measuring illuminance in the indoor of office and educational buildings. Jundishapur Journal of Health Sciences. 2014;6(3).
13. Kralikova R, Wessely E. Lighting Quality, Productivity And Human Health. Annals of DAAAM & Proceedings. 2016;27.
14. Pena-Garcia A. Towards total lighting: Expanding the frontiers of sustainable development. Multidisciplinary Digital Publishing Institute; 2019.
15. Rossler KL, Ganesh Sankaranarayanan D, Duvall A. Acquisition of fire safety knowledge and skills with virtual reality simulation. Nurse educator. 2019;44(2):88.
16. Schroll RC. Industrial fire protection handbook: CRC press; 2016.

17. Hurley MJ, Gottuk DT, Hall Jr JR, Harada K, Kuligowski ED, Puchovsky M Watts Jr JM, Wieczorek C J. SFPE handbook of fire protection engineering: Springer; 2015.
18. Wehmeier G, Mitropetros K. Fire protection in the chemical industry. Chemical engineering transactions. 2016;48:259–64.
19. Jaafar MH, Arifin K, Aiyub K, Razman MR, Ishak MIS, Samsurijan MS. Occupational safety and health management in the construction industry: a review. International Journal of Occupational Safety and Ergonomics. 2018;24(4):493–506.
20. Bhowmik D, Durai Vel S, Rajalakshmi A, Sampath Kumar K. Recent Trends in Hazards in the Pharmaceutical Industry and Safety Precaution. Elixir Pharmacy. 2014;69:23688–91.
21. Crow DR, Liggett DP, Scott MA. Changing the electrical safety culture. IEEE Transactions on Industry Applications. 2017;54(1):808–14.
22. Gore P, Mane A. Electrical Hazards And Safety. International Journal of Pure and Applied Mathematics. 2018;120(6):11997–2008.
23. Pinto DS, Clardy PF, Moreira ME. Environmental and weapon-related electrical injuries. Up to Date. 2016:1–112.
24. Zhao D, McCoy AP, Kleiner BM, Smith-Jackson TL. Control measures of electrical hazards: An analysis of construction industry. Safety Science. 2015;77:143–51.
25. Reese CD. Handbook of Safety and Health for the Service Industry-4 Volume Set: CRC Press. 2018.
26. Coco A, Jacklitsch B, Williams J, Kim J-H, Musolin K, Turner N. Criteria for a recommended standard: occupational exposure to heat and hot environments. control Ccfd, editor. 2016.
27. Lundgren K, Kuklane K, Venugopal V. Occupational heat stress and associated productivity loss estimation using the PHS model (ISO 7933): a case study from workplaces in Chennai, India. Global health action. 2014;7(1):25283.
28. Lucas RA, Epstein Y, Kjellstrom T. Excessive occupational heat exposure: a significant ergonomic challenge and health risk for current and future workers. Extreme physiology & medicine. 2014;3(1):1–8.
29. Gobir A, Aliyu A, Abubakar A, Ibrahim J, Esekhaigbe C, Joshua A, et al. Knowledge of heat waves and practice of protective measures against it in a rural West African. European Journal of Public Health. 2020;30(Supplement_5):ckaa166. 077.
30. Ncube F, Ncube EJ, Voyi K. Bioaerosols, noise, and ultraviolet radiation exposures for municipal solid waste handlers. Journal of environmental and public health. 2017.
31. Hudson HL, Nigam JA, Sauter SL, Chosewood L, Schill AL, Howard JE. Total worker health: American Psychological Association. 2019.
32. Little MP. Ionising radiation in the workplace. British Medical Journal Publishing Group; 2015.
33. Schulte PA, Delclos G, Felknor SA, Chosewood LC. Toward an expanded focus for occupational safety and health: a commentary. International journal of environmental research and public health. 2019;16(24):4946.
34. Jo Y-D, Ahn BJ. Analysis of hazard areas associated with high-pressure natural-gas pipelines. Journal of Loss Prevention in the Process industries. 2002;15(3):179–88.
35. Kent JA. Handbook of industrial chemistry and biotechnology: Springer Science & Business Media; 2013.
36. Scott RM. Chemical hazards in the workplace: CRC Press; 2020.
37. Asgedom AA, Bråtveit M, Moen BE. Knowledge, attitude and practice related to chemical hazards and personal protective equipment among particleboard workers in Ethiopia: a cross-sectional study. BMC public health. 2019;19(1):1–10.
38. Mansdorf SZ. Handbook of Occupational Safety and Health: Wiley Online Library; 2019.
39. Myrcha K, Gierasimiuk J. 18 Mechanical Hazards. Handbook of 7ccupational Safety and Health. 2010:359.
40. Holt ASJ, Allen J. Principles of health and safety at work: Routledge; 2015.
41. Chinniah Y. Analysis and prevention of serious and fatal accidents related to moving parts of machinery. Safety science. 2015;75:163–73.
42. Maiti J, Ray PK. Industrial Safety Management. Springer; 2018.

43. Takahashi M. Tackling psychosocial hazards at work. Industrial health. 2017;55(1):1–2.
44. Brous E. Workplace violence. AJN The American Journal of Nursing. 2018;118(10):51–5.
45. Arbury S, Zankowski D, Lipscomb J, Hodgson M. Workplace violence training programs for health care workers: an analysis of program elements. Workplace health & safety. 2017;65(6):266–72.
46. Wu Y-CJ, Wu T. A decade of entrepreneurship education in the Asia Pacific for future directions in theory and practice. Management Decision. 2017.
47. Maguire BJ, O'Neill BJ, O'Meara P, Browne M, Dealy MT. Preventing EMS workplace violence: a mixed-methods analysis of insights from assaulted medics. Injury. 2018;49(7):1258–65.

Chapter 13
Modified-Release Drug Delivery Systems

Saeed Ahmad Khan, Roohullah, and Alam Zeb

Abstract An ideal drug delivery system would have two prerequisites. Firstly, it should require a single dose for the entire treatment period (e.g., days or weeks in case of infection or for lifetime in case of hypertension). Secondly, it should deliver drug directly to the site of action. Obviously, such an imaginary delivery system is impossible. However, there has been enormous amount of research done over the years to eliminate fluctuation in plasma concentration observed with specified dosage regimen of conventional drug delivery systems. This chapter discusses the different approaches used for modifying the release of drug according to desired objective.

Keywords Sustained-release DDS · Monolithic systems · Osmotic pumps · Gastro-retentive DDS · Floating DDS · Mucoadhesive DDS

13.1 Introduction

There has always been interest in developing strategies to control the delivery of drugs, in order to improve the effectiveness of drugs. The objective in designing a controlled drug delivery system is to decrease dosing frequency, reduce required dose, localize at the site of action, or provide uniform drug delivery [1].

S. A. Khan (✉)
Department of Pharmacy, Kohat University of Science and Technology, Kohat, Pakistan
e-mail: saeedkhan@kust.edu.pk

Roohullah
Department of Pharmacy, Abasyn University Peshawar, Peshawar, Pakistan

A. Zeb
Riphah Institute of Pharmaceutical Sciences, Riphah International University, Rawalpindi, Islamabad, Pakistan

© The Author(s), under exclusive license to Springer Nature Switzerland AG 2022
S. A. Khan (ed.), *Essentials of Industrial Pharmacy*, AAPS Advances in the Pharmaceutical Sciences Series 46, https://doi.org/10.1007/978-3-030-84977-1_13

The delivery systems in which the drug release in terms of time course and/or location is modified to achieve therapeutic objective not achievable by conventional delivery systems are termed as modified-release (MR) dosage forms (according to USP).

13.2 Types of Modified-Release Dosage Forms

Worth to mention that USP considers that several terms have interchangeably been used in a confusing manner for different types of MR products, such as prolonged-release, sustained-release, and controlled-release systems. For the purpose of understanding, we classify all these products into three groups; sustained release drug delivery systems, Gastro-retentive drug delivery systems and controlled-release drug delivery systems

Sustained-Release Drug Delivery These dosage forms provide initial therapeutic dose promptly and then slowly release for longer time. Generally, the objective is to maintain therapeutic concentration of drug for an extended period. *Extended delivery* and *prolonged delivery* are also aimed at achieving sustained-release drug delivery. Contrarily, *repeat-action* systems contain multiple doses of drug in a unit dosage form. Therapeutic concentration is maintained by releasing a specified dose at a certain time [2].

Gastro-retentive Drug Delivery These systems rely on extending the gastric emptying time of delivery system, e.g., floating on the surface of gastric fluid or attaching to the mucus surface of the stomach, etc. [3].

Controlled-Release Drug Delivery The system attempts to control drug concentration at the site of action. In some cases, a controlled delivery system may not sustain the release of drug but may deliver the drug to a specific site. For example, *delayed-release* systems such as *enteric-coated* products or *colon-specific* products can be classified in controlled-release systems, because they delay the release of drug until they reach the intestine and colon, respectively [4]. Moreover, *site-specific delivery* and *targeted drug delivery* are descriptive terms used in the context of novel controlled-release drug carriers [5], that is discussed in detail, in Chap. 14

13.2.1 Sustained-Release Drug Delivery Systems

The complications in marketing new drugs have been increased for the last few decades. There are many diseases for which therapeutically effective compounds already exist, but these compounds are not effective in clinical settings because of the dose size or dosing frequency. For conventional peroral dosage forms to be

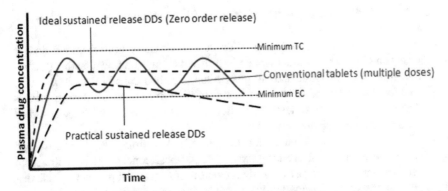

Fig. 13.1 Plasma drug concentration versus time for conventional tablet or capsule, ideal zero-order release systems, and practical first-order release systems

effective, a "steady state" plasma concentration needs to be achieved and maintained. Hence, a proper dose and correct dosing frequency is required. However, there are many potential limitations associated with conventional peroral delivery systems. For instance, even if "steady state" is achieved, the drug concentration in blood and hence at the site of action fluctuates in consecutive dosing interval [6]. Therefore, it is impossible to keep a constant drug concentration at the site of action for the duration of treatment. Moreover, the fluctuation can increase or decrease plasma drug concentration beyond the therapeutic range. Hence, it can lead to over- or undermedication of the patient. Furthermore, forgotten dose or overnight no-dose period has consequences on maintenance of therapeutic plasma concentration. This is more likely in cases of drugs with short half-lives, which need frequent dosing to maintain "steady state" plasma concentration [7], as shown in Fig. 13.1.

13.2.1.1 Advantages of Sustained-Release Drug Delivery System

- Improved control over maintenance of plasma drug concentration is helpful for treatment of diseases where symptoms subside as long as the drug concentration is greater than minimum effective concentration, e.g., asthma, depression, etc.
- Maintenance of therapeutic action in no-dose period, e.g., overnight pain relief for improved sleep.
- Avoid unwanted side effects related to very high peak plasma concentration [5].
- Reduce exposure of body to drug during total period of treatment, hence reducing side effects.
- Reduce dosing frequency, hence improving patient compliance.
- Reduce localized gastrointestinal irritation produced by irritant drugs, e.g., potassium chloride. That is why, potassium chloride is mostly delivered using MR form [7].
- Better disease management can reduce the total cost of treatment [8].

13.2.1.2   Limitations of Sustained-Release Dosage Forms

- Alike conventional peroral dosage forms, the release of drug from SR dosage forms and hence bioavailability can be affected by physiological factors (gastro-intestinal pH, enzymatic activity, gastric and intestinal transit time, etc.).
- The maximum period of therapeutic response of drug from SR dosage form is dependent on its gastrointestinal transit time and the time for which the absorbed drug is therapeutically active [5].
- SR products may stuck intact in some part of the gastrointestinal tract. In such cases, slow release of the drug may increase local concentration of drug and cause irritation of the gastrointestinal mucosa. However, SR products which disperse in the gastrointestinal fluids do not have such problems [5].
- There are limitations on the type of drugs that can be formulated as SR. For instance, drugs with very short half-lives, i.e., less than 2 hours, are very difficult to be formulated as SR product. Due to very high rate of elimination, an extremely large maintenance dose is required to provide continuous therapy for 8–12 hours. Very high dose is not only difficult to swallow but also a potential hazard [8].
- SR products usually contain larger doses than that administered with conventional dosage form. There is a potential hazard of overdosage if an improperly made SR product releases the whole amount at once or very fast. Therefore, SR product for very potent drugs is not recommended [9].
- For SR product to provide satisfactory prolonged therapy, drug needs to be absorbed from all regions of the gastrointestinal tract. Therefore, drugs having specific requirements for absorption from the GIT are poor candidates for SR dosage forms [8].
- Generally, per unit dose cost of SR formulations is more than conventional dosage forms, However, the total cost of treatment might be less since fewer "unit doses" of SR formulation are required.

13.2.1.3   Factors Affecting Design Strategy of SR Products

While designing sustained-release dosage form, it is necessary to consider various factors that can affect the drug delivery aspect of dosage form. Some of the important factors are:

(a) Biological properties of drug
(b) Physicochemical properties of drug

Biological Properties of Drug   Half-Life of Drug

The ultimate goal of sustained-release dosage form is to maintain therapeutic concentration of drug in blood for longer time. This can be achieved when the rate at which drug enters blood is equal to the rate at which it is eliminated. Drug elimination is sum of all processes which removes the drug from blood (e.g., metabolism, urinary excretion, etc.) [9]. The rate of elimination can be depicted from the half-life ($t_{1/2}$).

Drugs with shorter $t_{1/2}$ are ideal candidates for sustained-release systems; however drugs with very short half-life, i.e., less than 2 hours (e.g., furosemide or levodopa), may need extremely bigger doses. Therefore, these drugs are not suitable for sustained-release systems. Moreover, drugs with half-lives more than 8 hours (e.g., digoxin, warfarin, phenytoin, etc.) are also not recommended for sustained release, since they have intrinsic sustained-release effect [9]. Furthermore, it is difficult to prolong the duration of absorption beyond 8–12 hours, since the gastrointestinal transit time of most dosage forms is 8–12 hours [8].

Absorption of Drug

The absorption property of drugs also determines whether the drug is suitable for sustained-release dosage form. For compounds which have uniform rate of absorption throughout the gastrointestinal tract, the rate of drug release must be faster than the rate of absorption. Otherwise, the device will slip away the region of absorption before the drug is completely released [7]. It should also be noted that drugs with lower absorption rates have intrinsic sustained action and hence are not suitable for sustained-release dosage form [6].

However, for some drugs absorption may be limited to a specific region of the intestine, e.g., ferrous sulfate is primarily absorbed in the upper jejunum and duodenum. Therefore, sustained-release system may be not recommended for these drugs, since the delivery system may cross that specific site before releasing the drug. Instead, a recommended approach would be to maintain the drug in the stomach. This will release the drug slowly which will be then absorbed from the region of absorption. This was attempted by floating delivery systems. These delivery systems float and retain on the surface of gastric fluid, thus increasing gastric emptying time of drug [3]. Another suitable approach for these drugs is to use bioadhesive delivery system, which may adhere to the gastric surface, providing slow release of drug [1]. However, no substantial success has been reported for the use of oral bioadhesive delivery systems, though they have shown promises for use in the oral cavity, eye, and vagina.

Metabolism of Drug

Some drugs may be metabolized in lumen or intestinal tissue before absorption into bloodstream. Most of the metabolic enzymes in intestinal wall are saturable. If drug is released slowly, less drug is available to these enzymes, thus favoring more complete metabolism of drugs by these enzymes. Therefore, sustained-release delivery of these drugs is disadvantageous [2]. For example, oral delivery of alprenolol in a sustained-release delivery system showed extensive metabolism in the intestinal wall. Formulation of prodrugs for these drugs can be another suitable approach [10].

Physicochemical Factors *Dose Size*

For orally administered product (conventional as well as sustained release), the limit for a single dose is 0.5–1.0 g. Compounds requiring larger doses are usually given in multiple doses or in liquid dosage form [1]. However, there is a potential hazard of overdosage; therefore, very high doses are not recommended for MR products [9].

Solubility and Permeability

The aqueous solubility and intestinal permeability are of great importance while designing a sustained-release dosage from. Amidon et al. (1995) classified drugs into four categories on the basis of solubility and permeability (BCS classification system, discussed in chapter).

- BCS Class 1 drugs have high solubility and good permeability (ideal for sustained-release system).
- BCS Class 2 drugs have poor solubility and high permeability (not recommended for sustained release).
- BCS Class 3 drugs have high solubility and low permeability (the problem of low permeability).
- BCS Class 4, low solubility and low permeability (worst case for sustained release) [10].

Drugs having high solubility at intestinal pH and absorbed readily from the entire GIT by passive diffusion (i.e., not site-specific absorption) are excellent candidates for sustained-release dosage form [6]. Drugs having low solubility (less than 0.1 mg/ml) have an inherent sustained-release property, since their release over time will be limited due to their low solubility and, hence, are not recommended for sustained-release dosage from. The worst case of drugs is the ones having low solubility and low permeability, because the inherent sustaining action (due to low solubility) is not advantageous if the drug has low membrane permeability [8, 9].

Ionization

It should also be noted that most drugs are weak acids or bases; thus their ionization is dependent on pH of the surrounding. Since the uncharged form of the drug has higher permeability compared to ionized form, the unionized form of drug may seem advantageous for better absorption [7]. Unfortunately, the situation is not so simple. By changing to unionized form, the drug's aqueous solubility is lowered. Therefore, it is important to note the relationship of pKa and absorption environment before designing a sustained-release delivery system, because these devices have to function at different environments, from acidic (stomach) to neutral (small intestine). The effect of pH on release profile must be defined [11].

Partition Coefficient

Drug after administration to GI tract has to cross many biological membranes to reach the desired area and exert effect. It is understood that these membranes are usually hydrophobic; therefore partition coefficient of drugs is a critical factor for its permeation to the membrane. Compounds with high partition coefficient are lipophilic and have low water solubility [2]. Furthermore, lipophilic drugs remain in the body for longer time, since they may be distributed to the adipose tissues, e.g., phenothiazines. Partitioning effect is also important for diffusion of drug through polymeric membrane; therefore, it should be considered while choosing the type of release-limiting membrane [8].

13.2.1.4 Formulation of Sustained-Release Delivery Systems

As mentioned previously, in designing sustained-release oral delivery system, one would require rapid release of initial dose followed by slow release of the maintenance dose. For this reason, all sustained-release products have a physical "barrier" that restrains the release of maintenance dose. These barriers can be in the form of membranes, coatings, wax matrix, and polymer matrix, ion exchange resins and osmotic pumps [6]. For the rapid release of initial dose, a portion of drug is incorporated in rapidly releasing form, for example, as uncoated granules in tablet or capsule or placing a portion of drug on surface of wax or polymer matrix.

The components of sustained-release delivery systems are generally the same, though the types of excipients and method of incorporation may be different. These components include:

– Active ingredient (drug)
– Release-restraining agent (coating, matrix formers, membrane, ion exchange resin, etc.)
– Matrix or coating modifier (channeling agents for wax matrix and solubilizers for hydrophilic matrices)
– Other excipients to impart special characteristic (such as pH modifier, density modifiers, lubricants, bulking agent, etc.) [9]

13.2.1.5 Types of Sustained-Release Delivery System

There are many different types of sustained-release delivery systems [8, 9]. However, for convenience of understanding, these systems can be categorized on the basis of governing principle, such as:

– Matrix systems or monolithic systems
– Membrane-controlled (reservoirs) systems
– Osmotic pumps
– Ion exchange resins

Monolithic or Matrix Systems In matrix devices the drug is homogeneously dispersed in polymeric matrix. The polymeric matrix may be of two types:

Hydrophilic (Swellable/Soluble) Polymeric Matrix

These matrices swell when in contact with water forming a hydrated matrix layer. The rate of drug release is controlled by the diffusion of drug through the hydrated matrix layer. The outer swollen layer becomes more diluted with time and ultimately dissolves. These matrices act as a loose cross-linked polymer network that provides porous structure for drug diffusion [12].

Hydrophilic colloids are used as matrix forming material, such as sodium carboxymethyl cellulose, xanthan gum, alginate, etc.

For example, Brufen Retard®, ibuprofen extended-release tablet (Abbott), uses xanthan gum and hydroxypropyl methylcellulose as matrix former.

Isoptin SR®, verapamil hydrochloride sustained-release tablet (Abbott), uses sodium alginate and hydroxypropyl methylcellulose as matrix former [5].

Insoluble Polymeric Matrix

The drug is evenly distributed in a hydrophobic matrix containing channeling agent. The matrix remains intact in release medium while channeling agent gets dissolved. Leaching out of the dissolved channeling agent leaves capillaries in the matrix. Thus the fluid penetrates the matrix through these capillaries and dissolves the incorporated drug, which then diffuses out of the matrix. The release rate of drug from insoluble matrices depends on the convolution of the capillaries in the matrix, i.e., its porous structure [12]. The type of pore-forming hydrophilic salts and solutes (channeling agent) will greatly affect the porous structure of matrix. Generally, a rigid less porous matrix will show slower drug release, while a less consolidated highly porous matrix will show faster drug release.

Usually, insoluble polymers and insoluble lipid are used as matrix former.

For example, Procan SR®, procainamide hydrochloride extended-release tablet (Parke-Davis), uses carnauba wax as matrix material.

Ferrograd Folic®, ferrous sulfate and folic acid extended-release tablet (Abbott), uses insoluble methyl methacrylate copolymer as matrix former.

The drug release from matrices can be assessed using different mathematical models [13]. The most widely used equation is Korsmeyer-Peppas equation (Eq. 13.1):

$$M_t / M_\alpha = kt^n$$

(13.1)

where M_t/M_α is the drug fraction release at time t and k is the release rate constant and n is the release exponent that depicts the drug release mechanism of the device.

If the value of exponent n is 1.0, the drug release rate is independent of time, which means zero-order release kinetics. Here, relaxation and erosion of polymer are primarily responsible for drug release. If the value of $n = 0.5$, the release is primarily based on diffusion (Fickian). Values of n between 0.5 and 1 show that both diffusion and polymer relaxation contribute in the release kinetics (non-Fickian or anomalous or first-order release) [13].

Reservoir (Membrane-Controlled) Drug Delivery Systems As the name indicates, these are characterized by a core of drug (the reservoir) surrounded by a polymeric membrane through which the drug diffuses. It should be noted that the membrane does not swell nor it erodes. However, it has to become permeable, to allow diffusion of drug, e.g., through hydration by gastrointestinal fluid [12].

Reservoir systems can be either single-unit system, in which the whole unit (e.g., tablet) is coated with release-limiting membrane, or it can be multiple-unit system, in which individual particles (e.g., pellets) are coated. In any case the rate of drug release depends on the nature of membrane [8].

For example, Nico-400®, sustained-release niacin.

Osmotic Pump Systems The driving force for drug release from these systems is osmotic pressure; that is why they are called osmotic pumps. Typically, a semipermeable membrane surrounds the drug core. When the device comes in contact with gastrointestinal fluid, water is driven toward the core because of the higher osmotic pressure difference [15]. The rate of inward flow of water dV/dt through the semipermeable membrane can be represented by Eq. 13.2:

$$\frac{dV}{dt} = \frac{Ak}{h(\Delta\pi - \Delta P)}$$

(13.2)

where A = area of membrane, k = membrane permeability, h = membrane thickness, $\Delta\pi$ = osmotic pressure difference, and ΔP = hydrostatic pressure.

The rate of drug release is governed by the rate of inward flow of water through the membrane and the speed at which the drug can pass out of the hole [12].

These devices are typically of two types. The first contains a solid core containing drug and electrolyte. The electrolyte provides high osmotic pressure difference. The water flows inward because of osmotic pressure difference and dissolves the drug in the core. The drug is subsequently pushed out because of high hydrostatic pressure, through a hole in the coating [14].

The second type contains drug solution in an impermeable capsule surrounded by an electrolyte layer. The water driven inward due to osmotic pressure difference compresses the inner collapsible capsule containing drug, pumping the drug out of the hole. Alza Corporation's OROS® technology was one of the first devices of this type. This technology can be seen in other approved products, such as Acutrim®, Ditropan®, Glucotrol®, etc. [14].

An OROS® capsule contains a rigid semipermeable membrane coating having a laser-drilled orifice (diameter of 0.5–1.4 mm) for drug release (Fig. 13.2). The osmotically active polymer attracts the water through semipermeable membrane and swells pushing the drug out of the capsule through the orifice [15].

Ion-Exchange System Ion-exchange systems (IES) are generally insoluble polyelectrolytes having acidic or basic functional groups, capable of exchanging counterions in the gastrointestinal fluid. IES are usually in the form of beads prepared from organic polymer backbone. In these systems exchange of similar-charged ions of takes place. For instance, the drug loaded onto resin is released by exchanging with appropriately charged ions and then diffused to the gastrointestinal fluid [12].

The most common resins used in IES are cross-linked polystyrene and polymethacrylate polymers. They are of two types depending on type of functional groups

Fig. 13.2 Diagrammatic representation of osmotic pump systems

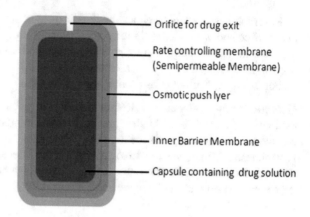

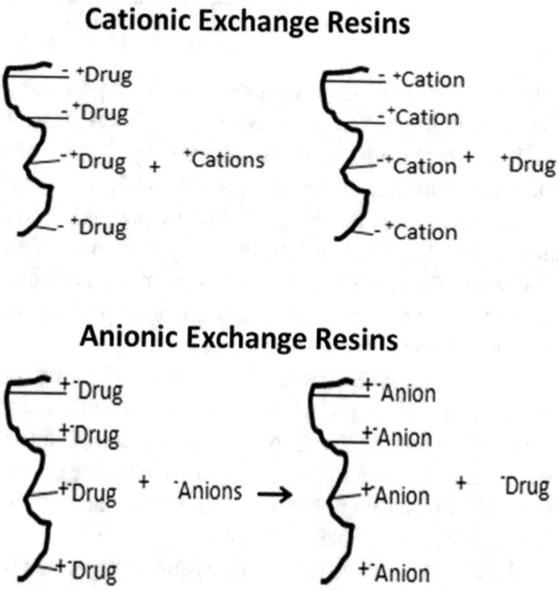

- Orifice for drug exit
- Rate controlling membrane (Semipermeable Membrane)
- Osmotic push lyer
- Inner Barrier Membrane
- Capsule containing drug solution

Fig. 13.3 Schematic representation of ion-exchange systems

Cationic Exchange Resins

{ –: +Drug
  –: +Drug
  –+Drug   +   +Cations
  –+Drug }

{ –: +Cation
  –: +Cation
  –+Cation   +   +Drug
  –+Cation }

Anionic Exchange Resins

{ +–Drug
  +–Drug
  +–Drug   +   –Anions →
  –+–Drug }

{ +–Anion
  +–Anion
  +–Anion   +   –Drug
  +–Anion }

they have. *Cationic ion-exchange resins* contain negatively charged functional groups (e.g., sulfonic acid groups, $-SO_3H$) and can exchange positively charged molecules and hence can be used for cationic drugs. *Anionic exchange resins* have positively charged functional groups (e.g., tertiary amines such as triethylamine) and are able to exchange negatively charged ions [16]. The mechanism of ion-exchange process is schematically shown in Fig. 13.3.

Some of the important factors affecting drug loading and release property of ion-exchange resins are:

- Area for diffusion and diffusional path length (depends on the size of the resin system)
- Rigidity of the resin (depends on degree of cross-linking)

– Available capacity for counterions (depends on number of ionic sites per unit weight or volume)
– Ionic strength (depends on the strength of acidic or basic functional groups) [16]

Ionamin® capsule (Phentermine, by Pennwalt) is an example of commercially available IES using cationic exchange resin.

13.2.2 Gastro-retentive Drug Delivery Systems

Oral route is one of the most widely used routes for administration of drug due to better patient compliance, acceptability, and ease of manufacturing. However, the oral route is not without its challenges. One major constraint in oral dosage form is their short gastric emptying time (GET). This can result in incomplete drug release and in turn less absorption in the zone of absorption (stomach or upper part of small intestine), thus decreasing the bioavailability and efficacy of drug. It is understood that the drug being stayed in the stomach for longer time may provide a sustained action of the drug. These types of formulations have prolonged gastric retention, hence known as gastro-retentive drug delivery systems (GRDDs) [3].

Before discussing the different approaches for GRDDs, it is better to revise physiology of gastric emptying.

13.2.2.1 Physiology of Stomach and Gastric Emptying

The stomach is composed of four parts (Fig. 13.4), the cardia, fundus, body, and pylorus:

Cardia the part surrounding the upper opening of the stomach
Fundus the rounded portion above and beside the cardia
Body the large central portion bellow the fundus
Pyloric antrum and pyloric canal the part connecting the stomach to the duodenum

The process of gastric emptying occurs through an interdigestive series of electrical events occurring in the stomach, called interdigestive myoelectric cycle or migrating myoelectric cycle (MMC). This occurs in four stages:

Phase 1 (basic phase) lasts from 30 to 60 min with rare contractions.
Phase 2 (preburst phase) lasts for 20–40 min with fluctuated contractions, and the frequency and intensity of contraction increases as the phase progresses.
Phase 3 (burst phase) lasts for 4–6 min and is a short period of regular intense contractions. It is also termed as "housekeeper wave," since it sweeps the undigested material toward the intestine.
Phase 4 is a short (0–5 min) transition phase between phase 3 and phase 1 of two consecutive cycles.

Fig. 13.4 Diagrammatic representation of different regions of the stomach

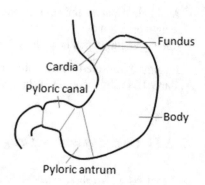

MMC occurs in both the fasting and fed states. However, feeding results in delaying the onset of MMC; hence the rate of gastric emptying is reduced in fed state [17].

13.2.2.2 Types of Gastro-retentive DDS

Over the last two decades, several strategies have been employed for designing gastro-retentive dosage forms to prolong gastric residence time [3, 17]. Some of the common types are given as follows:

(a) Sinking (high-density) systems
(b) Floating (low-density) systems
(c) Mucoadhesive systems
(d) Magnetic systems

Sinking (High-Density) Systems The density of gastric contents is almost similar to that of water (1.004 g/cm³). In upright position, small high-density objects may sink to the bottom of the stomach and lodge in the pyloric antrum (Fig. 13.5a) and survive the peristaltic waves of the stomach wall. Usually, a density of around 2.5 g/cm³ is necessary for any object to sink in the stomach [17]. High-density excipients such as barium sulfate, zinc oxide, and iron powder are used to impart high density to the delivery system. Although encouraging results have been reported for high-density delivery systems, however, no such product has reached to the market yet [18].

Floating Drug Delivery Systems (FDDs) The density of these delivery systems is less than gastric fluids, thus remaining buoyant (float) on the surface of gastric fluid (Fig. 13.5b) for longer time and continuously releasing the drug for longer time. Floating drug delivery systems must serve as reservoir and slowly release drug. Moreover, the density must be lower than that of gastric contents. The density of delivery system can be decreased by air entrapment (e.g., hollow chambers), by adding low-density materials such as oils or by adding gas generating material (Effervescent Systems) [3].

Fig. 13.5 Representation of high-density drug delivery systems (**a**) and floating drug delivery systems (**b**)

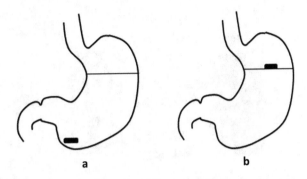

FDDs can be in the form of single unit (e.g., tablet) or multiple units (e.g., microspheres, beads, etc.). FDDS can be classified into two groups on the bases of mechanism of buoyancy: non-effervescent systems and effervescent systems.

Non-effervescent Systems

Non-effervescent FDDS are usually prepared from highly swellable polymers. The air entrapped in the swollen polymer network confers floating property to these delivery systems. The common examples of non-effervescent systems are hydrodynamically balanced systems and hollow microspheres/microballoons.

Hydrodynamically balanced systems were designed by Sheth and Tossounian in 1984. These are single-unit dosage forms, containing drug incorporated in one or more gel-forming hydrophilic polymers and administered in gelatin capsule. The capsule shell rapidly dissolves exposing hydrocolloid to gastric fluid. Hydration and swelling of the polymer produces a floating mass. The drug diffuses from the swollen polymer layer [18]. Continuous erosion of the polymer surface permits water to penetrate inward, thus maintaining surface hydration and floating (schematically shown in Fig. 13.6).

Hydroxypropyl methylcellulose (HPMC) is the most common hydrocolloid used in hydrodynamically balanced systems. Other polymers commonly employed are polyacrylates, polyvinyl acetate, agar, polyethylene oxide, carbopol, etc. The following are some of the commercial products utilizing HBS principle [17].

Madopar HBS® (levodopa and benserazide), an anti-Parkinson's agent by Hoffmann-La Roche, utilizes the principle of hydrodynamically balanced system. Valrelease® is another floating product, by Hoffmann-La Roche, containing diazepam. The delivery system forms a soft floating gelatinous mass in the stomach, hence maximizing diazepam absorption. Since diazepam is more soluble at low pH, the stomach is more favorable for its absorption than the intestine. Liquid Gaviscon® (contains aluminum hydroxide and magnesium carbonate in sodium alginate solution, by GlaxoSmithKline) forms a low-density gel "raft" that floats on the surface of gastric fluid [18].

Hollow microspheres/microballoons containing drug in their outer polymeric shell (Fig. 13.7) were prepared by Kawashima et al. (1992) to extend the gastric

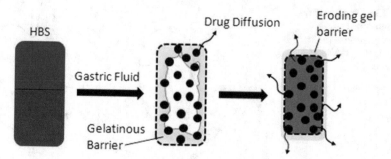

Fig. 13.6 Diagrammatic representation of hydrodynamically balanced systems

Fig. 13.7 Diagrammatic representation of microballoons

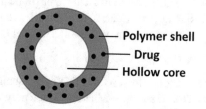

retention time (GRT) of drug. Polymers such as polycarbonate, cellulose acetate, calcium alginate, Eudragit S, etc. are used for preparation of hollow microspheres.

Floating and drug release form microballoons depend on various factors, some of which are very critical, such as types of polymers, concentration of polymer and drug to polymer ratio, types of plasticizers used, etc. [17].

Effervescent Systems

Effervescence (generation of gas bubbles) can also cause floatability. For this purpose, carbonates or bicarbonates are incorporated in the delivery system which generates CO_2 by reacting with acid (either co-incorporated citric or tartaric acid or natural gastric acid). For generation of gas, an optimum stoichiometric ratio of base to acid is needed, for example, 0.76:1 for citric acid and sodium bicarbonate.

Effervescence systems can be monolayer, bilayer, or multilayer systems (Fig. 13.8) in single-unit form (tablets) or multiple-unit form (microparticles). The gas generating components are incorporated in hydrophilic polymer, which upon swelling forms porous matrix. CO_2 is generated as a result of acid-base reaction (shown in Eq. 13.3) and is trapped in the porous swollen matrix, causing the delivery system to float on gastric fluid. These systems can further be modified by coating the matrix with a polymer that is impermeable to CO_2 but permeable to water.

$$C_6H_8O_7 + NaHCO_3 \rightarrow C_6H_8O_6Na + 3H_2O + 3CO_2$$

Citric acid   Sodium bicarbonate   Sodium citrate   Water   Carbon dioxide

(13.3)

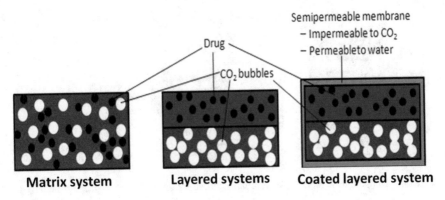

Fig. 13.8 Different types of effervescent delivery systems

Some of the examples of commercial products based on the principle of effervescence for floating are Prazopress XL® (Prazosin HCl, by Sun Pharma, Japan) and Oflin OD® (Ofloxacin, by Ranbaxy, India) [3].

Advantages of FDDs Floating dosage forms provide higher gastric retention time for drugs, consequently increasing the overall absorption of drugs. It is advantageous for drugs acting locally, such as antacids. Moreover, it is useful in case of diarrhea to keep the drug for longer time in GIT. Additionally, useful for drugs which can cause local irritation if released abruptly in larger amount, e.g., potassium chloride [18]. It is advantageous for drugs that are primarily absorbed from the stomach e.g., ferrous salts, antacids, diazepam, etc. It is also useful for drugs that are unstable in intestinal or colonic environment, e.g., captopril, ranitidine HCl, etc. Moreover, it is helpful for drugs that show lower solubility at alkaline pH, e.g., diazepam, chlordiazepoxide, and verapamil HCl [3].

Limitations of FDDS Not suitable for drugs that have limited acid solubility, e.g., phenytoin, etc. Additionally, FDDS is not suitable for drugs that are selectively intended for action in the colon, e.g., 5-aminosalicylic acid, corticosteroids, etc. Moreover, it is not suitable for drugs that are unstable in gastric fluids, e.g., erythromycin. Furthermore, it is not suitable for drugs which undergo significant first-pass metabolism, e.g., estrogens, enalapril, nifedipine, etc. FDDS require a sufficient gastric fluid for efficient floating [17].

Mucoadhesive or Bioadhesive Systems The concept of mucoadhesive drug delivery devices has gained huge interest since 1980s. The American Society of Testing and Materials (ASTM) defines adhesion as the state where two surfaces are held together by interfacial forces (valence forces and/or interlocking action).

Mucoadhesive drug delivery system adheres to the mucous membrane (Fig. 13.9) (a thin moist membrane covering various body cavities), consequently, increasing the residence time of dosage form at the site of application and thus increasing the therapeutic performance of drug [1].

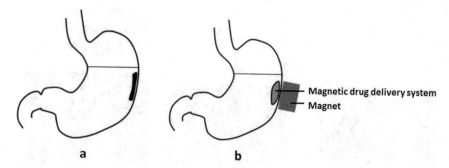

Fig. 13.9 Diagrammatic representation of mucoadhesive systems (**a**) and magnetic systems (**b**)

A number of mucoadhesive drug delivery devices have been designed for local or systemic effects, through buccal, nasal, rectal, and vaginal routes. Recently, its use as gastro-retentive dosage form has also shown great promises.

Commonly used polymers for mucoadhesive delivery systems are chitosan, poly(methacrylic acid), sodium carboxymethyl cellulose, poly-acrylic acid, etc. [18].

Magnetic Systems These systems contain drug and magnetically active compounds such as iron oxide (Fe_2O_3). The delivery system is held in the stomach with the help of a magnet, i.e., a magnet is placed on the abdomen over the position of the stomach, as shown in Fig. 13.9. Although these systems have shown success for single-unit system as well as multiple-unit systems, requirement of precision in application of external magnet may compromise patient compliance [17].

References

1. Bailey, MM., Berkland, CJ., (2010) Modified Release Delivery Systems. In: Morishita M., Park M., (Eds) Biodrug Delivery Systems. CRC Press, Boca Raton, Florida; pp 250–63. https://doi.org/10.3109/9781420086713-17
2. Fassihi, R. (2017) Modified-Release Delivery Systems (2017). In: Augsburger, BL., Hoag, SW., (Eds) Pharmaceutical Dosage Forms. CRC Press, Boca Raton, Florida; pp 317–44. https://doi.org/10.1201/9781315111896-12
3. Zaman, M., Akhtar, F., Baseer, A., Hasan, SMF., Aman, W., Khan, A., Badshah, M., Ullah, M (2021). Formulation development and in-vitro evaluation of gastroretentive drug delivery system of loxoprofen sodium: A natural excipients based approach. Pakistan Journal of Pharmaceutical Sciences. 34 (1): p. 057-063.
4. Siegel, RA., Rathbone, MJ (2012). Overview of Controlled Release Mechanisms. Fundamentals and Applications of Controlled Release Drug Delivery. Springer US 27;19–43. doi.org/10.1007/978-1-4614-0881-9_2
5. Majumder, J., Taratula, O., Minko, T, (2019). Nanocarrier-based systems for targeted and site specific therapeutic delivery. Advanced Drug Delivery Reviews;144:57–77. doi.org/10.1016/j.addr.2019.07.010

6. Singh, L., Sharma, V. (2021). Implementation of Quality by Design principles for the evolution of optimized sustained release drug delivery system. Drug Delivery Letters; 11(3): 233-247. doi.org/10.2174/2210303111666210421121812
7. Hasnain, MS., Ray, P., Nayak, AK. (2020). Alginate-based interpenetrating polymer networks for sustained drug release. In: Nayak, AK., and Hasnain, MS. Alginates in Drug Delivery. Elsevier, Amsterdam, Netherlands; pp.101–28. doi.org/10.1016/b978-0-12-817640-5.00005-4
8. Nashed, N., Lam, M. and Nokhodchi, Ali.,(2021). A comprehensive overview of extended release oral dosage forms manufactured through hot melt extrusion and its combination with 3D printing. International Journal of Pharmaceutics: 596, 120237. doi.org/10.1016/j.ijpharm.2021.120237
9. Abdellatif, A. (2018). Design of Sustained Action Dosage Forms; Mini-Review. Pharmaceutica Analytica Acta: 09(07);1000593. https://doi.org/10.4172/2153-2435.1000593
10. Papich, MG. and Martinez MN. (2015). Applying Biopharmaceutical Classification System (BCS) Criteria to Predict Oral Absorption of Drugs in Dogs: Challenges and Pitfalls; American Association of Pharmaceutical Scientists Journal; 17(4): 948–964. doi: 10.1208/s12248-015-9743-7
11. Inamdar, SN., Ahmed, K., Rohman, N., Skelton, AA. (2018). Novel pKa/DFT-Based Theoretical Model for Predicting the Drug Loading and Release of a pH-Responsive Drug Delivery System. American Chemical Society;122 (23), 12279–90 https://doi.org/10.1021/acs.jpcc.8b02794.s001
12. Nigusse, B., Gebre-Mariam, T., and Belete, A., (2021). Design, development and optimization of sustained release floating, bioadhesive and swellable matrix tablet of ranitidine hydrochloride. PLoS ONE 16(6): e0253391.https://doi.org/10.1371/journal.pone.0253391
13. Ullah, M., Ullah, H., Murtaza, G., Mahmood, Q., Hussain, I. (2015) Evaluation of Influence of Various Polymers on Dissolution and Phase Behavior of Carbamazepine-Succinic Acid Cocrystal in Matrix Tablets. BioMed Research International:2015:1–10. https://doi.org/10.1155/2015/870656
14. Chen, J., Pan, H., Ye, T., Liu, D., Li, Q., Chen, F, et al.(2016). Recent Aspects of Osmotic Pump Systems: Functionalization, Clinical use and Advanced Imaging Technology. Current Drug Metabolism:17(3):279–91. https://doi.org/10.2174/1389200216666151015115706
15. Conley, R., Gupta, SK., Sathyan, G. (2006). Clinical spectrum of the osmotic-controlled release oral delivery system (OROS), an advanced oral delivery form. Current Medical Research and Opinion:22(10);1879-1892. https://doi.org/10.1185/030079906x132613.
16. Guo, X., Chang, RK., Hussain, MA. (2009). Ion-exchange resins as drug delivery carriers, Journal of Pharmaceutical Sciences;98(11):3886-3902. https://doi.org/10.1002/jps.21706.
17. Tripathi, J., Thapa, P., Maharjan, R., Jeong SH. (2019). Current State and Future Perspectives on Gastroretentive Drug Delivery Systems. Pharmaceutics: 20;11(4):193. https://doi.org/10.3390/pharmaceutics11040193
18. Pund, AU., Shendge, RS., Pote, AK. (2020). Current Approaches on Gastroretentive Drug Delivery systems. Journal of Drug Delivery and Therapeutics: 10(1):139–46. https://doi.org/10.22270/jddt.v10i1.3803

Chapter 14
Novel Drug Delivery Systems

Saeed Ahmad Khan and Hussain Ali

Abstract Novel drug delivery typically involves nanotechnology-based approaches that require multidisciplinary efforts combining the fields of pharmaceutics, physical chemistry, polymer science, bioconjugate chemistry, and molecular biology. This chapter discusses various aspects of novel drug delivery. The first part of the chapter is especially focused on targeted drug delivery approaches, such as passive targeting and active targeting. The later part of the chapter provides valuable insight into various nanopharmaceuticals employed as novel drug delivery systems: drug nanocrystals, polymeric nanoparticles, liposomes, and niosomes are discussed as examples.

Keywords Targeted drug delivery · Nanomedicine · Nanoparticles · Pharmaceutical nanotechnology · Nanocrystal technology

14.1   Introduction

Novel drug delivery system (NDDS) refers to the delivery systems that impart unique therapeutic properties to the drug that are otherwise not exhibited by these drugs in raw form. These unique properties may be in terms of improved bioavailability, patient compliance, safety, and efficacy, etc.

Nanotechnology is the technology that deals with development of materials within nanometer range. The concept for application of nanotechnology, to design safer and effective drug delivery and diagnostics, was visualized more than 40 years ago. However, this goal became seemingly achievable in recent years with many

S. A. Khan (✉)
Department of Pharmacy, Kohat University of Science and Technology, Kohat, Pakistan
e-mail: saeedkhan@kust.edu.pk

H. Ali
Department of Pharmacy, Quaid-I-Azam University Islamabad, Islamabad, Pakistan

© The Author(s), under exclusive license to Springer Nature Switzerland AG 2022   235
S. A. Khan (ed.), *Essentials of Industrial Pharmacy*, AAPS Advances in the
Pharmaceutical Sciences Series 46, https://doi.org/10.1007/978-3-030-84977-1_14

products commercialized for treatment and prevention of many critical diseases, such as cancer, cardiovascular diseases, HIV/AIDS, arthritis, etc.

Novel drug delivery typically employs nanotechnology-based approaches to design drug particles, drug-carrier particles, or drug-carrier complexes, in the size range of 10–1000 nm (nanopharmaceuticals) to improve solubility, increase stability, improve permeability, reduce toxicity, and increase accumulation of drug in desired tissue. Targeted drug delivery is one of the most common examples that novel drug delivery systems are aimed at.

14.2 Targeted Drug Delivery

The term targeted drug delivery is sometimes interchangeably used with *site-specific delivery* and *controlled drug delivery*. However, the term *targeted drug delivery* is more relevant in the context of NDDS, since the expression *controlled drug delivery* is more appropriate for conventional drug delivery that provides a controlled release of drugs in the intestine without necessarily prolonging the release of drug [1]. For example, drug delivery exhibited by *delayed-release systems*, such as *enteric-coated systems* or *colon-specific systems*, delays the release of drug until they reach the intestine and colon, respectively. These delivery systems are not meant to sustain the release of drug but rather aimed at delivering the drug to specific site in the gastrointestinal tract.

This section is primarily focusing on targeted drug delivery that aims at delivering the drug specifically to the target site for improving therapeutic efficiency, reducing toxicity, or preventing harmful side effects, particularly in cancer, since conventional cancer therapeutic strategies (surgery, radiation, and chemotherapies) have limitations. There are two approaches for targeted drug delivery, i.e., passive targeting and active targeting.

14.2.1 Passive Targeting

Passive targeting refers to accumulation of drug at the target site by exploiting the pathophysiology of the disease, without any attached targeting ligand [2]. The microenvironment of the diseased tissue differs from normal tissue in many ways, such as vascular abnormalities, oxygenation, perfusion, pH, etc. This phenomenon is exploited for passive targeting.

Selective accumulation of nanocarriers by passive targeting mainly relies on the phenomenon called *enhanced permeation and retention (EPR)*, since EPR effect can be observed in almost all types of cancer except hypovascular cancers such as pancreatic cancer and prostate cancer. Hence, EPR effect has become gold standard in passive targeting.

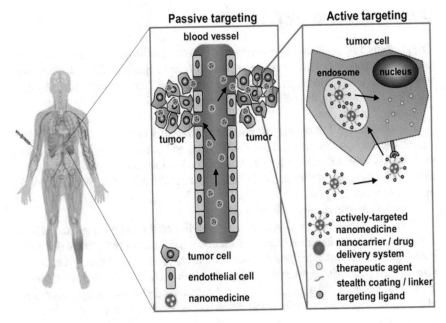

Fig. 14.1 Mechanism of active and passive targeted drug delivery to tumors [7]

14.2.1.1 EPR Effect

EPR effect is a combination of two phenomena, i.e., enhanced permeation and enhanced retention, as shown in Fig. 14.1.

Enhanced Permeation In cancer the cells grow rapidly and abnormally. The normal vasculature is unable to meet the nutritional requirement of rapidly growing tissue. This leads to angiogenesis (creation of new blood vessels from existing ones). The newly formed vessels have irregular endothelial layer (due to presence of high proportion of proliferating endothelial cells), abnormal basement membrane, and deficiency of pericyte (contractile cells surrounding the endothelial cells). This results into a leaky vasculature with larger pores (10–1000 nm). Consequently, substances of sufficiently smaller size can readily pass through the pores into tumorous interstitium.

Enhanced Retention In normal tissue lymphatic systems are responsible for the drainage. The lymphatic vessels carry excess of the substances that are leaked into the interstitium. In tumor tissue these lymphatic vessels are either absent or damaged. Consequently, substances that have entered extracellular fluid are not removed efficiently, hence retained in tumor tissue.

14.2.1.2   Factors Affecting EPR Effect

The permeation of molecules through blood vessels and the movement of molecules across extracellular fluid are affected by many factors. Some of the prominent factors include tumor vasculature, tumor extravascular environment, and physicochemical properties of the colloids.

Tumor Vasculature

The secretion of angiogenic factors, such as vascular permeability factor (VPF), vascular endothelial growth factor (VEGF), etc., and in turn the leakiness of tumor vasculature, vary with cancer type and its stages. For instance, murine colon adenocarcinoma is hyperpermeable, while human pancreatic adenocarcinoma is poorly permeable tumors [3, 4].

Tumor Extravascular Environment

Normally, interstitial fluid is efficiently distributed throughout the extracellular environment. While in tumor, the extravascular environment is congested due to the presence of many entangled proteins (i.e., collagen, laminin, etc.) and glycosaminoglycans (i.e., heparin, hyaluronic acid, chondroitin sulfate, etc.). The nature of extracellular matrix is dependent on protein content and glycosaminoglycan molecular weight, which in turn affects the diffusion/convection of nanocarriers in tumor interstitium.

Physicochemical Properties of Colloids

Size   The permeation of colloids through tumor vasculature and its accumulation at tumor site are greatly affected by size of the colloid. Reports have shown that smaller colloids (2–3 nm) permeate faster and diffuse deeper in the tumor tissue, as compared to larger colloids (7–25 nm) [5].

Charge   The charge on colloids influences the interaction of colloids with vascular endothelial cells as well as with other charged macromolecules present in extracellular matrix, e.g., collagen, heparin, hyaluronic acid, etc. Consequently, the diffusion of colloids through blood vessels and movement in the interstitium is greatly affected by the charge of colloid.

Shape   The extravasation and tumor distribution kinetic are also affected by geometry of the colloid. Studies conducted on nanorods and nanospheres of same diameter have shown that nanorods permeate four times faster, than nanospheres, through the tumor vasculature. Moreover, nanorods diffuse deeper in extracellular matrix compared to nanospheres. These results suggest that rod-shaped colloids might be beneficial in terms of better exploitation of EPR effect [6].

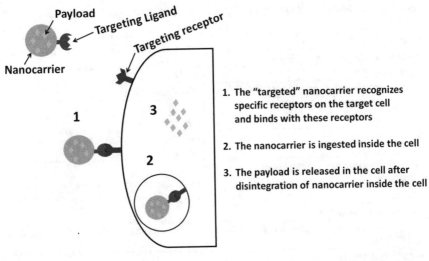

Fig. 14.2 Schematic representation of ligand-receptor pair for active targeted drug delivery

14.2.2 Active Targeting

Active targeting (also known as ligand-mediated targeting) involves guiding the drug-loaded carrier specifically toward specific cells, as schematically shown in Fig.14.2. This is achieved by attaching receptor-specific molecules (*targeting ligands*) to the drug carrier, which not only guides the drug carrier toward target site but also increases interaction of drug carrier with the cells, thus improving cellular uptake. The receptors that are exploited for active targeting are:

- The markers that are present in normal cells and are overexpressed in tumor cells due to metabolic abnormalities, e.g., folate receptor (FR) and transferrin receptor (TfR)
- New markers on the tumor cells that appear due to the genetic defects in tumor cells

The different types of targeting ligand and corresponding receptors are given in Table 14.1.

14.3 Nanopharmaceuticals

Nanopharmaceuticals are drug particles, drug-polymer particles, or drug-polymer complex in the size range of 10–1000 nm that are employed to improve solubility, increase stability, improve permeability, reduce toxicity, and/or increase accumulation of drug in desired tissue, that are otherwise not exhibited by these drugs in raw form.

Table 14.1 Examples of targeting ligands and the corresponding receptors used for targeted drug delivery [8]

Targeting ligands	Corresponding receptor
Proteins and peptide-based ligands	
Transferrin (Tf)	Tf receptor
Vascular endothelial growth factor (VEGF)	VEGF receptor
Vasoactive intestinal peptide (VIP)	VIP receptors
Somatostatin	Somatostatin receptor
Angiopep-2	Low-density lipoprotein receptor-related protein 1 (LRP1)
Monoclonal antibody or fragments	
Cetuximab	Epidermal growth factor receptor (EGFR)
Bevacizumab	Vascular endothelial growth factor (VEGF)
Nucleic acid-based ligands	
2-[3-(1,3-Dicarboxypropyl)ureido] pentanedioic acid (DUPA)	Prostate-specific membrane antigen (PSMA)
Sugar-based ligands	
β-D-glucose	Glucose transfer 1 (Glut 1) receptor
Lectin-based ligands	
Wheat germ agglutinin (WGA)	N-Acetylglucosamine (GlcNAc)
Small molecules	
Folic acid	Folate receptor

According to the report published in 2015 [9], the total market size for nano-medicine was expected to be about 1 trillion US dollar. With progressive growth rate in the research and more efficient treatment strategies in various diseases, the marketplace all over the world is expected to show a compound annual growth rate (CAGR) of 22% for the nanoparticles in biomedical sciences [10].

Nanopharmaceuticals are generally classified in many ways, such as chemical composition, size, shape, and method of synthesis. From material perspective, nanopharmaceuticals are of three types:

(a) Organic nanoparticles
(b) Inorganic nanoparticles
(c) Carbon-based nanoparticles

14.3.1 Organic Nanoparticles

Organic nanoparticles are further classified into drug nanoparticles, drug-loaded polymeric nanoparticles, and drug-loaded nanovesicles.

14.3.1.1 Drug Nanoparticles

When the solid core of nanoparticles is entirely composed of drug, these are termed as *drug nanoparticles*, which may be either in the form of nanocrystal or nanoamorphous, depending on the molecular state of drug. Nevertheless, nanocrystals are more common compared to nanoamorphous systems. Therefore, the following section will primarily focus on drug nanocrystals.

Drug Nanocrystals These are atom aggregates that assemble into a "cluster" of nanometer dimension [11]. By definition, nanocrystal dimensions are less than 1000 nm; however, for practical purposes, size less than 500 nm is usually aimed for [12, 13]. Drug nanocrystals are typically stabilized by a thin surfactant layer [14].

Preparation of Drug Nanocrystals

Manufacturing and stabilization of nanocrystals is an expensive technology. Moreover, not all drugs can comply with nanocrystal technology [15]. The approaches for preparation of drug nanocrystals are of two types, i.e., *top-down approach* and *bottom-up approach* [16].

Top-Down Approach

Top-down approach is also called *nanonization*. It involves size reduction of coarse drug particles into nanometer range, using milling or high-pressure homogenization (HPH).

Milling A modified ball mill (as discussed in Section 3.7.1.3) is used for nanonization. Beads of around 1 cm diameter, made up of zirconium oxide, glass, or stainless steel, are used for milling. By impact and attrition mechanism of the beads, coarse drug particles are reduced to nanometer size range. Milling does not require organic solvents; hence it is eco-friendly. Moreover, milling is performed in close system; therefore, it reduces the chances of contamination.

High-Pressure Homogenization (HPH) A preformed drug slurry is forced through a narrow orifice at high pressure (discussed in detail in Sect. 3.7.2.4). The particle size is reduced by the phenomenon of turbulence, shear stress, and cavitation caused by the movement of fluid though a narrow orifice and impact, as shown in Fig. 14.3.

The key drawback associated with HPH is that it may affect the crystalline nature of nanocrystals. Moreover, contamination due to long period of homogenization and generation of high amount of energy are also unwanted issues.

Bottom-Up Approach

Bottom-up approach is also called *crystal growth* or *nucleation*. This technique relies on self-assembly of primary building blocks (molecules) into a relatively bigger structure (nanocrystal). Some of the advantages of this technique include simple equipment, low energy requirement, lower heat generation, and low cost. Bottom-up technique is equally effective for thermostable and thermolabile drugs. One of the

Fig. 14.3 Basic
mechanism of particle size
reduction by high-pressure
homogenization

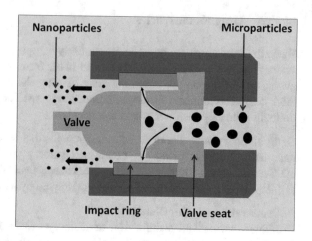

most common bottom-up strategies for nanocrystal preparation is antisolvent precipitation.

Antisolvent Precipitation It is one of the most common examples of bottom-up approach. The drug is dissolved in suitable solvent and it is subjected to an antisolvent; consequently the drug molecules assemble to form crystals. However, this aggregation of molecules needs to be controlled with the help of stabilizers, such as polymers, surfactants, etc.

Application of Drug Nanocrystals

Nanocrystal technology in pharmaceutical development is primarily used to increase the solubility of drugs. An example of the effect of particle size reduction on drug dissolution is given in Fig. 14.4.

The drug nanocrystal technology is well established. There has been immense commercial success of nanocrystals as nanopharmaceuticals, lately. Some of the commercial products are summarized in Table 14.2.

14.3.1.2 Drug-Loaded Polymeric Nanoparticles

Polymeric nanoparticles are comprised of solid matrix that can encapsulate a variety of molecules, such as drugs, biologicals, or diagnostic agents. The drug is encapsulated inside the core, adsorbed physically, or linked covalently to the surface of nanoparticles. The first polymeric nanoparticles were poly(acrylamide)-based micelles that were introduced in the 1970s for encapsulation of tetanus toxoid or human IgG [18]. However, the inverse microemulsion polymerization technique for synthesis of nanoparticles was obsoleted later, since the presence of residual reactants and initiators posed serious toxicity concerns.

Polymers Used for Nanoparticles

Various biodegradable and biocompatible polymers have been used for fabrication of nanoparticles. These polymers may be from synthetic origin, such as

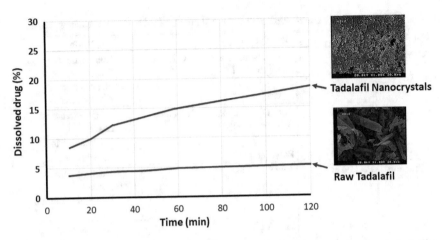

Fig. 14.4 Dissolution profiles of raw tadalafil and tadalafil nanocrystals in phosphate buffer (pH = 7.4) medium at 37.0 ± 0.5 °C. Each value represents the mean ± SD (n = 3) (adopted from Teymouri et al., [17])

Table 14.2 Examples of commercial products based on nanocrystal technology

Drug name	Trade name/company	Nanonization technology used	Ultimate dosage form
Sirolimus	Rapamune®/Pfizer	Top-down, milling	Tablet
Fenofibrate	TriCor®/AbbVie	Top-down, milling	Tablet
Aprepitant	Emend®/MSD	Top-down, milling	Capsules and injection
Fenofibrate	Triglide®/Skyepharma	Top-down, HPH	Tablet
Megestrol acetate	Megace ES®/Par Pharmaceutical	Top-down, milling	Oral suspension
Griseofulvin	Gris-PEG®/Valeant	Bottom-up, precipitation	Tablet
Nabilone	Cesamet/Meda	Bottom-up, precipitation	Capsule

poly(lactide) (PLA), poly(glycolic acid) (PGA), or copolymer of PLA and PLGA, i.e., poly(lactide-co-glycolide) (PLGA), polycaprolactone (PCL), polyethylene glycol (PEG), and polyurethane. Moreover, natural polymers, such as albumin, gelatin, alginate etc., have also been showing great promise. In any case, the polymer employed for nanoparticle synthesis need to be biodegradable, so that it is fully metabolized into monomers and subsequently excreted from the body. Examples of commercially available drug-loaded polymeric nanoparticles are given in Table 14.3.

14.3.1.3 Drug-Loaded Nanovesicles

Nanovesicles are a type of organic nanoparticles composed of an aqueous core surrounded by a bilayer membrane. These vesicles are formed by self-assembly of amphiphilic molecules, such as phospholipids, nonionic surfactants, and

Table 14.3 Examples of commercial products based on drug-loaded polymeric nanoparticles

Name	Drug	Description	Application
Accurins™	Docetaxel	Poly(DL-lactic acid)-PEG block copolymer NPs	Lung cancer Prostate cancer Head cancer Neck cancer
Genexol®	Paclitaxel	Poly(DL-lactic acid)-PEG block copolymer NPs	Metastatic breast cancer Pancreatic cancer
Opaxio®	Paclitaxel	Polyglutamate NPs	Glioblastoma
Abraxane®	Paclitaxel	Albumin NPs	Metastatic breast cancer Non-small cell lung cancer
Eligard®	Leuprolide acetate	PLGA NPs	Prostate cancer

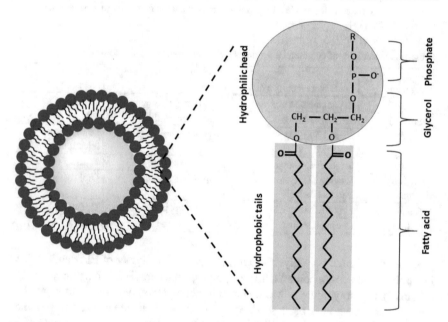

Fig. 14.5 Diagrammatic representation of liposomes and basic structure of phospholipids

amphiphilic block copolymers; the vesicles formed are called liposomes, niosomes, and polymersomes, respectively.

Liposomes

The word liposome derives from two Greek words: *Lipo* means "fat" and *soma* means "body." Liposomes are spherical vesicles that are made up of phospholipids [19]. The phospholipids are arranged in a bilayer structure having hydrophobic tails directed inward while polar heads outward, as shown in Fig. 14.5.

The physicochemical properties of liposomes and rigidity vary with the bilayer components and phospholipids used [20]. For instance, unsaturated phospholipids, such as soybean phosphatidylcholine, form permeable liposomes and less stable bilayer, whereas unsaturated phospholipids with long acyl chains, such as dipalmitoylphosphatidylcholine, form liposomes that are rigid and impermeable.

Phospholipids are esters of fatty acids, phosphoric acids, glycerol, and other alcohols, for example, phosphatidylcholine, phosphatidylethanolamine, phosphatidylinositol, and phosphatidylserine. These all consist of fatty acids and esters at one or two positions of the glycerol molecule with phosphate group esterified at position three.

The first-generation liposomes, composed of phospholipids and cholesterol, could reduce toxicity and improve therapeutic index of entrapped drug. However, these liposomes were prone to rapid engulfment by reticuloendothelial system (RES). In order to circumvent this issue, stealth liposomes are designed by coating liposome with polyethylene glycol, i.e., PEGylated liposomes. PEGylated liposomes have increased circulation time and increased stability of drug by non-engulfment [21], thus maximizing bioavailability of drug.

A variety of therapeutic agents can be loaded in liposomes, such as anticancer drugs, biologicals, and antimicrobials. The hydrophilic molecules are typically loaded into the aqueous core, while lipophilic substances are loaded in the bilayer membrane of liposome [22]. The payload can also be physically or covalently attached to the surface of liposomes. Table.14.4 represents some of the examples of commercially available liposomal products for drug delivery.

Niosomes

Niosomes are lamellar structures (nonionic surfactant vesicles) formed by combination of nonionic surfactant with cholesterol [23]. Niosomes are similar to liposomes in terms of morphology; however, instead of phospholipids, nonionic surfactant is used in niosomes. Consequently, the physicochemical properties of niosomes differ from that of liposomes, as summarized in Table 14.5.

The major components of niosomes are nonionic surfactant and cholesterol; surfactant is responsible for formation of lamellar structure, while cholesterol provides rigidity and stability to the shape of niosome. The nature of surfactants influences encapsulation efficiency, toxicity, and stability of niosomes. Nonionic surfactants of the Span® series and Tween® series are generally used for the preparation of niosomes [24]. The properties of niosomes, such as size, surface charge, lamellarity, and trapped volume, can be varied by changing the composition. For instance, incorporation of dicetyl phosphate, phosphatidic acid etc., in the membrane can impart negative charge, while stearylamine, cetylpyridinium chloride, etc. can form positively charged niosomes [24]. The surfactant/lipid and surfactant/water ratios are critical considerations for formulation designing, since encapsulation efficiency of niosomes is greatly affected by these parameters [25]. A surfactant/lipid ratio of 1/2.5 w/w is typically used for formation of stable niosomes.

Table 14.4 Examples of liposomal products available for clinical applications

Brand name	Drug	Indication
Ambisome™	Amphotericin B	Fungal infection
DaunoXome™	Daunorubicin	Kaposi's sarcoma
Doxil™	Doxorubicin	Ovarian cancer, AIDS-related Kaposi's sarcoma, multiple myeloma
Visudyne®	Verteporfin	Age-related macular degeneration, pathologic myopia, and ocular histoplasmosis
Myocet®	Doxorubicin	Recurrent breast cancer
DepoCyt®	Cytarabine	Neoplastic meningitis and lymphomatous meningitis
Lipoplatin®	Cisplatin	Epithelial malignancies
DepoDur®	Morphine sulfate	Postoperative pain following major surgery

Table 14.5 Some of the differences between niosomes and liposomes

Niosomes	Liposomes
– Primarily composed of nonionic single-chain surfactant and cholesterol	– Prepared from neutral or charged double-chain phospholipids
– Economical and stable compared to liposomes	– Phospholipids are expensive and chemically unstable. Therefore, liposomes require special storage condition

14.3.2 Inorganic Nanoparticles

A variety of inorganic materials have been investigated for biomedical applications. For some of the inorganic materials, e.g., gold, silver, iron oxide, quantum dots, etc., nanonization imparts unique electronic properties compared to their parent bulk material. These properties have been extensively exploited for diagnosis and therapeutic purposes.

Gold nanoparticles and iron oxide nanoparticles are two examples of inorganic nanoparticles that have a long history of use in biomedical field. Therefore, these nanoparticles will be explained in detail in the coming section.

14.3.2.1 Gold Nanoparticles

Gold nanoparticles (AuNPs) have centuries-old history in chemistry and biological sciences [26]. The experiment proposed by Michael Faraday, around 150 years ago, has been considered a baseline for the modern era of AuNP synthesis. Moreover, in 1971, the British researchers Faulk and Taylor [27] developed antibody conjugation of salmonella antigen with colloidal gold surface coating for direct electron microscopy visualization. Since then, the use of colloidal gold has been greatly increased in different scientific domains such as chemistry, biology, engineering, and medicines. Lately, AuNP has been applied to a variety of medical applications, for example, gene delivery, radiation therapy, and diagnosis, etc.

The presence of plasmon absorbance bands and their shape- and size-dependent properties make gold NPs a versatile material for diverse applications in many fields (as summarized in Table 14.6).

For biomedical applications, Aurasol® was the first product based on AuNPs that was developed for the management of rheumatoid arthritis. However, it is no more used. The antiangiogenic property of AuNPs was later exploited for the development of Aurimune® that has recently completed phase 1 trials and has shown good tolerability and tumor deposition. The paclitaxel-loaded Aurimune® platform is under preclinical development stage. However, the only AuNP-based product approved for tumor elimination to date is AuroLase® that uses silica nanoparticles coated with gold nanoparticles.

14.3.2.2 Iron Oxide Nanoparticles (IONPs)

For the past 90 years, materials such as iron oxides, iron hydroxide, or their combination have been extensively used in biomedical field due to their unique properties, such as magnetic properties, superparamagnetic properties, and biodegradability. IONPs have been used in iron replacement therapy, magnetic resonance imaging (MRI), biosensors, targeted drug delivery, and cell separation [29].

When the size of iron oxide materials is reduced to nanometer range, it exhibits superparamagnetic properties. These particles are referred to as superparamagnetic iron oxide NPs (SPIONs) [30]. In this connection maghemite (γ-Fe_2O_3), hematite, (α-Fe_2O_3) and magnetite (Fe_3O_4) have proven experimentally to be highly magnetizable when external magnetic field is applied. However, SPIONs tend to agglomerate in magnetic field; therefore, effective surface stabilization is required. Typically, fatty acids, PEG, poly(N-vinylpyrrolidone), dextran, chitosan, etc. are used as coating agents [31].

The early parenteral formulation of colloidal iron oxide for anemia was stabilized with dextran. However, due to rapid release of elemental iron, the formulation raised toxicity concerns [32]. Therefore, thick coating with carbohydrate was introduced to mitigate the rapid release of elemental iron in the circulation, e.g., Imferon® (Fisons) or Dexferrum®. Later on, low molecular weight dextran, sucrose,

Table 14.6 Application of AuNPs on the basis of their properties, adopted from [28]

Properties	Major field of application
Redox activity	Electrochemical sensing and electronic devices
Surface-enhanced Raman scattering (SERS)	Imaging and sensing
Surface plasmon resonance (SPR)	Colorimetric analysis and photothermal therapy
Fluorescence quenching	Sensing and material fabrication

gluconate, etc. were introduced for stabilization of iron oxide formulations, and consequently, the products were less toxic and exhibited little immunogenic reaction [33].

Modern intravenous iron formulations are iron oxide/hydroxide nanoparticles stabilized with carbohydrates [34]. A variety of iron oxide-based formulation are available for treatment of anemia, such as Ferinject® and Injectafer®, containing ferric carboxymaltose, which is commercialized by Vifor Pharma and Daiichi Sankyo. Other examples are Venofer® containing iron-sucrose complex manufactured by Vifor Pharma, Monofer® containing iron isomaltoside marketed by Pharmacosmos in Europe, and Feraheme®/Rienso® containing ferumoxytol as SPIONs, manufactured by AMAG Pharmaceuticals.

Additionally, SPIONs have been used as contrast agent for MRI. For example, Feridex® and Endorem® containing iron oxide stabilized with dextran are used for imaging of liver and spleen lesions. Other examples include Resovist® and Cliavist® containing ferucarbotran with carboxydextran, marketed by Bayer Healthcare, used for imaging of liver lesions. Additionally, Ferrotran®, previously knowns as Combidex based on Ferumoxtran-10, is the only MRI contrast agent that can detect lymph node metastases as small as 2 mm [35].

Besides being used as a diagnostic agent, SPIONs have also recommended for hyperthermia treatment of cancer. Since SPIONs have high magnetic saturation, therefore it is also employed in the target site to damage cells locally by increasing temperature. Based on the aforementioned principles, a new product is introduced into the market known as NanoTherm® (MagForce), which is also available in Europe since 2018. These tiny magnetic nanoparticles are introduced either locally into tumor, which is then heated by external magnetic field, named as NanoActivator®, which destroys the cancer cells.

References

1. Heng, P.W., Controlled release drug delivery systems. Pharmaceutical Development and Technology, 2018. **23**(9): p. 833-833.
2. Kang, H., et al., Size-dependent EPR effect of polymeric nanoparticles on tumor targeting. Advanced Healthcare Materials, 2020. **9**(1): p. 1901223.
3. Cabral, H., et al., Accumulation of sub-100 nm polymeric micelles in poorly permeable tumours depends on size. Nature Nanotechnology, 2011. **6**(12): p. 815-823.
4. Maeda, H., The enhanced permeability and retention (EPR) effect in tumor vasculature: the key role of tumor-selective macromolecular drug targeting. Advances in Enzyme Regulation, 2001. **41**(1): p. 189-207.
5. Dreher, M.R., et al., Tumor Vascular Permeability, Accumulation, and Penetration of Macromolecular Drug Carriers. Journal of the National Cancer Institute, 2006. **98**(5): p. 335-344.
6. Chauhan, V.P., et al., Fluorescent Nanorods and Nanospheres for Real-Time In Vivo Probing of Nanoparticle Shape-Dependent Tumor Penetration. Angewandte Chemie International Edition, 2011. **50**(48): p. 11417-11420.

7. Danhier, F., Feron, O and Préat, V. To exploit the tumor microenvironment: passive and active tumor targeting of nanocarriers for anti-cancer drug delivery. Journal of Controlled Release, 2010. **148**(2): p. 135-146.

8. Bertrand, N., et al., Cancer nanotechnology: the impact of passive and active targeting in the era of modern cancer biology. Advanced Drug Delivery Reviews, 2014. **66**: p. 2-25.

9. ETPN – Nanomedicine European Technology Platform. Src, Ncpm, Ncrd. Strategic agenda for EuroNanoMed. (accessed 28 March 2022) *https://etp-nanomedicine.eu/about-nanomedicine/ strategic-research-and-innovation-agenda/.*

10. Global Markets for Nanoparticle Size Analysis Instrumentation in the Life Sciences. (accessed 28 March 2022). https://www.bccresearch.com/market-research/biotechnology/nanoparticle-size-analysis-instrumentation-life-sciences-report.html

11. Viswanathan, P., Muralidaran, Y and Ragavan, G. (2017). Challenges in oral drug delivery: a nano-based strategy to overcome, In: Andronescu, E, Grumezescu, A., (Eds) Nanostructures for oral medicine. 2017, Elsevier, Amsterdam, Netherlands: p. 173–201.

12. Chavda, V.P., (2019). Nanobased nano drug delivery: a comprehensive review. In: Applications of Targeted Nano-Drugs and Delivery Systems: Nanoscience and Nanotechnology in Drug Delivery, Elsevier Inc., Amsterdam, Netherlands: p. 69-92.

13. Newton, A.M.J and Kaur, S, (2019), Solid lipid nanoparticles for skin and drug delivery: Methods of preparation and characterization techniques and applications. In: Grumezescu, AM., Nanoarchitectonics in Biomedicine. 2019, Elsevier, Amsterdam, Netherlands: p. 295-334.

14. Pund, S. and Joshi, A., Nanoarchitectures for neglected tropical protozoal diseases: challenges and state of the art, in Nano-and Microscale Drug Delivery Systems. 2017, Elsevier, Amsterdam, Netherlands: p. 439-480.

15. Saini, J.K. and Kumar, S. Development of nanocrystal formulation with improved dissolution. Journal of Drug Delivery and Therapeutics, 2018. **8**(5): p. 118-129.

16. Mohammad, I.S., et al., Drug nanocrystals: fabrication methods and promising therapeutic applications. International Journal of Pharmaceutics, 2019. **562**: p. 187-202.

17. Teymouri Rad, R., et al., Enhanced Dissolution Rate of Tadalafil Nanoparticles Prepared by Sonoprecipitation Technique: Optimization and Physicochemical Investigation. Iranian Journal of Pharmaceutical Research, 2017. **16**(4): p. 1335-1348.

18. Birrenbach, G. and Speiser, P.P., Polymerized micelles and their use as adjuvants in immunology. Journal of Pharmaceutical Sciences, 1976. **65**(12): p. 1763-6.

19. Sharma, V.K. and Agrawal, M.K. A historical perspective of liposomes-a bio nanomaterial. Materials Today: Proceedings, 2021. **45**: p. 2963-2966.

20. Roy, B., et al., Influence of lipid composition, pH, and temperature on physicochemical properties of liposomes with curcumin as model drug. Journal of oleo Science, 2016. **65**(5): p. 399-411.

21. Mohamed, M., et al., PEGylated liposomes: immunological responses. Science and Technology of Advanced Materials, 2019. **20**(1): p. 710-724.

22. Alavi, M., Karimi, N and Safaei, M., Application of various types of liposomes in drug delivery systems. Advanced Pharmaceutical Bulletin, 2017. **7**(1): p. 3.

23. Shah, A., et al., Advanced development of a non-ionic surfactant and cholesterol material based niosomal gel formulation for the topical delivery of anti-acne drugs. Materials Advances, 2020. **1**(6): p. 1763-1774.

24. Ge, X., et al., Advances of non-ionic surfactant vesicles (niosomes) and their application in drug delivery. Pharmaceutics, 2019. **11**(2): p. 55.

25. Abdelkader, H., Alani, A.W.G and Alany, R.G. Recent advances in non-ionic surfactant vesicles (niosomes): self-assembly, fabrication, characterization, drug delivery applications and limitations. Drug delivery, 2014. **21**(2): p. 87-100.

26. Dykman, L. and Khlebtsov, N. Gold nanoparticles in biology and medicine: recent advances and prospects. Acta Naturae, 2011. **3**(2 (9)).

27. Faulk, W.P. and Taylor, G.M. Communication to the editors: an immunocolloid method for the electron microscope. Immunochemistry, 1971. **8**(11): p. 1081-1083.
28. Yeh, Y.-C., Creran, B and Rotello, V.M., Gold nanoparticles: preparation, properties, and applications in bionanotechnology. Nanoscale, 2012. **4**(6): p. 1871-1880.
29. Arias, L.S., et al., Iron oxide nanoparticles for biomedical applications: a perspective on synthesis, drugs, antimicrobial activity, and toxicity. Antibiotics, 2018. **7**(2): p. 46.
30. Sharifi, S., et al., Superparamagnetic iron oxide nanoparticles for in vivo molecular and cellular imaging. Contrast Media & Molecular Imaging, 2015. **10**(5): p. 329-355.
31. Sodipo, B.K. and Aziz, A.A., Recent advances in synthesis and surface modification of superparamagnetic iron oxide nanoparticles with silica. Journal of Magnetism and Magnetic Materials, 2016. **416**: p. 275-291.
32. Goetsch, A.T., Moore, C.V., and Minnich, V. observations on the effect of massive doses of iron given intravenously to patients with hypochromic anemia. Blood, 1946. **1**(2): p. 129-142.
33. Jahn, M.R., et al., A comparative study of the physicochemical properties of iron isomaltoside 1000 (Monofer®), a new intravenous iron preparation and its clinical implications. European Journal of Pharmaceutics and Biopharmaceutics, 2011. **78**(3): p. 480-491.
34. Schaefer, B., et al., Intravenous iron supplementation therapy. Molecular Aspects of Medicine, 2020. **75**: p. 100862.
35. Bisso, S. and Leroux, J.-C., Nanopharmaceuticals: A focus on their clinical translatability. International Journal of Pharmaceutics, 2020. **578**: p. 119098.

Chapter 15
Polymer for Biomedical Applications

Iqra Riasat, Muhammad Naeem, Muhammad Umar Aslam Khan, Syed Babar Jamal, Atif Ali Khan Khalil, Sajjad Haider, and Adnan Haider

Abstract The structure of a polymer depends entirely on its monomer. Polymers are broadly classified into two main groups, natural and synthetic, depending on the source from which they are derived. Polymers have made a name for themselves in all areas of science and especially in the biomedical field due to their wide range of applications. Among the numerous polymers, biopolymers have attracted the attention of the scientific community mainly because of their biocompatible and biodegradable properties. In this chapter, we have summarized information about polymers, their classification, and last but not least, their potential biomedical applications. This will help students to understand the basic concepts related to polymers and especially biopolymers and their potential application in the biomedical field.

Keywords Polymeric scaffolds · Biocomposite · Biodegradable · Biocompatible

I. Riasat
Department of Bioinformatics and Biosciences, Capital University of Science and Technology (CUST), Islamabad, Pakistan

M. Naeem · S. B. Jamal · A. A. K. Khalil · A. Haider (✉)
Department of Biological Sciences, National University of Medical Sciences, Rawalpindi, Pakistan

M. U. A. Khan
Department of Polymer Engineering and Technology, University of the Punjab, Lahore, Pakistan

S. Haider (✉)
Department of Chemical Engineering, College of Engineering, King Saud University, PO-BOX 800, Riyadh, 11421, Saudi Arabia
e-mail: shaider@ksu.edu.sa

© The Author(s), under exclusive license to Springer Nature Switzerland AG 2022
S. A. Khan (ed.), *Essentials of Industrial Pharmacy*, AAPS Advances in the Pharmaceutical Sciences Series 46, https://doi.org/10.1007/978-3-030-84977-1_15

15.1   Introduction

A polymer is a large molecule made up of small repeating units, called *mono-mers. Monomers are formed into a large molecule by a process called polymeriza-tion*. Polymers (synthetic and natural) have a positive impact on daily life due to their numerous applications in various fields. The central role of polymers is evident in from the fact that everything from plastics to proteins and nucleic acids are com-prised of these natural and synthetic polymers. Polymerization is the process by which small repeating units are combined/linked in such a way that they stack up and form long-chain structures called polymers [1]. Due to their unique physical and chemical properties (tensile strength, viscoelasticity, biocompatibility, biode-gradability, ease of moulding), they are used in almost all fields of science [2, 3].

The word "polymer" is derived from the Greek words *polus* (means "many) and *meros* (means "part"), referring to molecules containing repeating units. Thus, a polymer is defined as a large molecule made up of small repeating units, called *monomers,* by a process called *polymerization* [1]. Jöns Jacob Berzelius in 1833 used the term although his definition was different from the IUPAC system. It was Hermann Staudinger in 1920 who from his findings proposed that polymers are covalently bonded macromolecules [4]. Recently, however, researchers have been exploring non-covalently bonded supramolecular polymers [5, 6].

The number of monomers, types of monomers, and their orientation in a polymer structure may vary from polymer to polymer. The properties of polymer can be tailored by the polymerization process. The simplest polymer is formed by bonding just the same type of monomer, and the polymer is called *homopolymer*. Similarly, when different types of monomers are combined, the resultant polymer is called het-ero and *copolymer*. Copolymers can vary in structure, composition, and the mono-mer ratio, which ultimately define the chemical and physical properties of the copolymer. The interlinking of polymer chains molecules to one another is called *cross-linking*, for instance, rubber bands and polyethylene bags.

15.2   Classification of Polymers

There are different types of polymer classifications. For instance, in 1929 Carothers classified polymers on the basis of structure: addition polymers and condensation polymer [7]. In another classification, polymers are simply categorized as linear polymers, branched polymers, or cross-linked polymers (Fig. 15.1).

In the context of this chapter, classification of polymer on the basis of degrad-ability will be more relevant: biodegradable polymers and nonbiodegradable poly-mers. However, here the focus is primarily on biodegradable polymers, since these polymers have more applications in drug delivery.

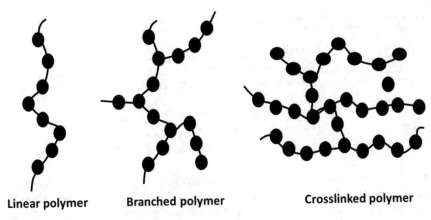

Linear polymer   Branched polymer   Crosslinked polymer

Fig. 15.1 Representation of polymer types based on the linkage of monomers, linear polymer, branched polymer, and cross-linked polymers

15.2.1   *Biodegradable Polymers*

Previously nonbiodegradable polymers were more common in the biomedical field. However, the problem with such polymers was that they would remain in the body forever and would need surgical removal after the desired therapeutic objective is achieved. This was the reason that scientists started focusing on biodegradable polymers, which when introduced to the body are degraded (by hydrolysis, enzymatic degradation, or combination of both).

Another critical property of polymers for biomedical use is biocompatibility. The polymer itself or its metabolites must not have toxic effects on the body [7].

Polymers for biomedical application are further classified on the basis of their origin; they are of two types: natural polymers and synthetic polymers.

15.2.1.1   Natural Biodegradable Polymers

Polymers that have a natural origin are known as the natural polymers. These polymers are obtained from plants, animals, or microorganisms and are also called *biopolymers*. These polymers include proteins, polynucleotides, and polysaccharides [5, 6]. Biopolymers received a lot of attention because of their beneficial properties such as biodegradability, lack of toxicity, availability, low cost, and, most importantly, biocompatibility.

Biopolymers have wide range of applications in various fields and more specifically in the pharmaceutical industry [8]. In the pharmaceutical industry, biopolymers are used as coating, gelling, emulsifying agents, binder, and disintegrants in capsule and tablet formulation.

Some of the common biopolymers are summarized in Table 15.1. One of the most abundant natural polymers is cellulose, which is obtained from plants and

Table 15.1 Representation of natural polymers based on their origin

Origin of natural polymers	Polymers
Animal	Gelatin, hyaluronan, chitin
Plant	Cellulose, starch, hemicellulose, agar, pectin, guar gum, psyllium, gum acacia, lignin
Microbe	Xanthan, gellan, hyaluronan, curdlan
Algae	Carrageenan, alginate, agar
Fungus	Schizophyllan, cardlan, scleroglucan, pullulan, chitin

mostly used in the laboratories for the fabrication of fibers, clothes, cosmetics, paper, pharmaceutical formulations, etc. Some of the semisynthetic forms of cellulose are also prepared in laboratories that include cellulose ethers and cellulose esters, which like the pristine cellulose have found application in preparation of fillers, binders, compressibility enhancers, and gelling agents [2, 9].

However, biopolymers show relatively higher susceptibility toward microbial contamination when they are exposed to the external environment [10–13]. Moreover, it is difficult to purify polymers once it gets contaminated with impurities during extraction [11].

15.2.1.2   Synthetic Biodegradable Polymers

Synthetic polymers are man-made and are considered a relatively new class of polymers, having been introduced only in the early twentieth century. These polymers are synthesized in the laboratory through chemical reactions that allows the fabrication of a wide variety of polymers [7, 14].

Biocompatible synthetic polymers offer a wide range of applications in the biomedical field, such as drug delivery, hemodialysis, tissue engineering, etc. Polymers with the ability to respond to certain internal or external stimuli, such as temperature, pH, and ionic strength, can be used in stimulus-responsive drug delivery systems [15].

Synthetic biodegradable polymers typically have unstable bond links in the backbone, usually a carbonyl bond attached to an oxygen, nitrogen, or sulfur atom (Fig. 15.2). Depending on the attached functional groups, a wide range of polymers have been synthesized that have shown potential for application in the biomedical field [16]. Some of the common classes of synthetic biodegradable polymers are given as follows.

Polyesters The commonly used polymers included polylactic acid (PLA) and polyglycolic acid (PGA) (structure shown in Fig. 15.3). These polymers were initially used for absorbable sutures.

PLA is an optically active biodegradable polymer that can be synthesized in the laboratory and derived from plant starch, corn, cassava, maize, sugarcane or

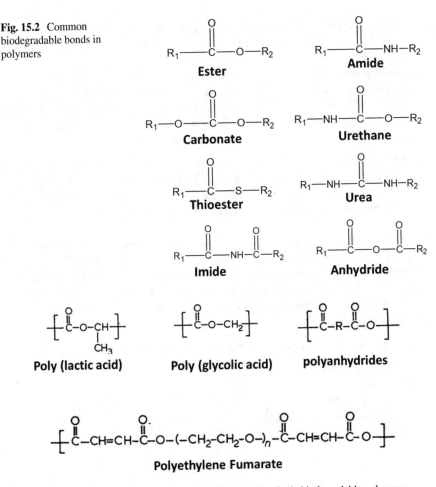

Fig. 15.2 Common biodegradable bonds in polymers

Fig. 15.3 General structure of the most common class of synthetic biodegradable polymers

sugar beet pulp. The L-lactic acid form is easily metabolized by the human body; therefore, it is preferred over D-form for drug delivery system. In pure polymeric form PLA is crystalline and not easily hydrolyzed. Whereas their racemic mixture is glassy and almost completely noncrystalline and hydrolyzes more rapidly. So racemic mixture is more frequently used for controlled-release drug delivery system. PLA degrades by hydrolysis, and the higher the molecular weight, the more time is required for hydrolysis.

PGA is highly crystalline; therefore, it hydrolyzes slowly. Hence, it is used in combination with PLA to control the degradation process (e.g., PLGA).

PLGA is a copolymer of PLA and PGA. The percentage ratio of PGA may be 0–30%. The higher the amount of PGA, the slower will be the degradation. Furthermore, the degradation process is also affected by the type of substitution near ester group. For instance, introducing an electron-donating group, such as alkyl or aryl group, decreases the hydrolysis, either by a steric hindrance effect or an electron donation that stabilizes the oxygen of the ester bond, or vice versa with an electron-withdrawing group such as an amine group.

Poly(ε-caprolactone) (PCL) is a polyester containing longer hydrocarbon chain in the polymer backbone, which results in a much lower glass transition temperature (Tg), i.e., −60 °C. PCL is semicrystalline; hence it exhibits slower degradation. Due to its rubbery nature, PCL is much more permeable than PLA and PLGA; therefore it exhibits much faster drug release.

Cross-linked Polyesters The typical example of cross-linked polyesters is formed by the reaction of diacids and a diol or polyol (polyethylene glycol, PEG), e.g., polyethylene fumarate (Fig. 15.3). The reactions yield a water-soluble copolymer that becomes insoluble, when cross-linked with N-polyvinylpyrrolidone.

Polyanhydrides Polyanhydrides are formed by the condensation of polyacids (Fig. 15.3). The anhydride linkage is highly hydrolytic, and a small change in the polymer backbone can lead to large changes in the physicochemical properties. For instance, the introduction of hydrophobic groups into the ring slows down hydrolysis, while the rate of degradation is increased by the incorporation of hydrophilic groups into the backbone.

Other less common synthetic biodegradable polymers include:

- Poly(alkylcyanoacrylates)
- Polyorthoesters
- Polyphosphazenes
- Poly(D-3-hydroxybutyric acid)
- Polydioxanones and polyoxalates

15.2.2 Application of Biodegradable Polymers

Biopolymers have mostly been used as carrier or supporting materials for drug delivery. Below are some of the applications of biopolymers in biomedical field [17].

Nanofibrous Scaffolds for Tissue Engineering The nano and micro features in addition to chemical structures play an important role in tissue engineering scaffolds [18, 19]. The fibrous scaffolds should provide a native environment to the cells [20]. Various methods have been adopted by researchers for the fabrication of nano/micro fibers, among these electrospinning is considered to be one of the most sophisticated technique. The fibers prepared through electrospinning have inter- and intra-fiber pores. These pores facilitate the penetration of cells into the scaffolds [21]. Similarly, a thermally induced phase separation technique has been developed to fabricate 3D PLLA nanofibrous scaffolds [22]. The fabricated scaffolds provide a large surface area, high porosity, and inter- and intra-fiber porosity that provide a native environment to the cells.

Injectable scaffolds based on biocompatible polymers have also been extensively investigated. They have an edge over the conventional approaches, since injectable scaffolds can be easily manipulated and require minimal invasion [23]. A group of researchers synthesized star-shaped poly(L-lactic acid) (SS-PLLA) and fabricated injectable hollow nanofibrous microspheres through a self-assembly technique [24]. The hollow nanofibrous microspheres mimicked the extracellular matrix (ECM) and provided a native environment to the cultured cells, which in turn contributed in the regeneration of cartilage. The hollow nanofibrous microspheres/chondrocytes also exhibited enhanced cartilage repair in a critical size rabbit model. The results suggested that hollow nanofibrous microsphere can be an excellent tool for cartilage repair [25].

Nanocomposite Scaffolds for Bone Tissue Regeneration Single-component polymeric scaffolds have been widely used in tissue engineering. However, using single/pristine polymers at times does not meet the entire criterion required for tissue regeneration. Bone matrix is a complex architecture as it consists of organic and inorganic materials such as collagen and apatite, respectively. Numerous studies have reported the preparation of scaffolds from biocomposite polymers containing apatite for use in bone tissue applications [26, 27]. Hydroxyapatite (HAP) $(Ca_{10}(PO_4)_6(OH)_2)$ is the most frequently used ceramic material in bone tissue regeneration scaffolds. Hydroxyapatite is mainly composed of calcium and phosphorus with a specific ratio and is used for enhancing osteogenesis. In other studies, bioglass and other combinations of calcium and phosphate have also been used for bone tissue regeneration [28]. Lei et al. reported that nanofibrous gelatin silica composite scaffolds, fabricated through a thermally induced phase separation (TIPS), can also be used for the aforementioned purpose [29].

Drug Delivery Systems Proteins and drugs are an ever-growing class of pharmaceutics which are used for the treatment of various diseases. Most drugs and proteins are target specific [30, 31]. With the advancement in technology, new methods have been introduced for delivering drugs to target site [2, 32]. In almost all the methods biodegradable and biocompatible polymers are used mainly as carriers which are capable of providing native environment to cells. In a study by Haider et al., pamidronic acid was grafted on the surface of hydroxyapatite nanorods. The pamidronic acid-grafted hydroxyapatite was incorporated into the PLGA polymer matrix. The pamidronic acid-loaded hydroxyapatite-PLGA composite nanofiber scaffolds were subjected to in vitro analysis. From the obtained data they concluded that pamidronic acid has reduced osteoclastic cell activity (bone-resorbing cells), whereas osteoblastic cell (bone-forming cells) activity was enhanced due to the presence of hydroxyapatite [33]. Similarly, they also performed experiments on elucidating the efficacy of PLGA composite scaffolds loaded with BMP-2 and insulin [13, 34, 35].

Furthermore, numerous methods have been implemented for fabricating biocompatible and biodegradable polymeric scaffolds loaded with drugs/proteins and nano-/microparticles. Among the methods, electrospinning, solvent casting, melt molding, freeze-drying, 3D printing, and salt leaching are the most commonly used methods to fabricate biodegradable polymeric scaffolds for biomedical applications and more specifically for drug delivery purposes [2, 6].

References

1. Post W, Susa A, Blaauw R, Molenveld K, Knoop RJI. A Review on the Potential and Limitations of Recyclable Thermosets for Structural Applications. Polymer Reviews. 2020;60(2):359-88.
2. Haider A, Haider S, Kummara MR, Kamal T, Alghyamah A-AA, Iftikhar FJ, et al. Advances in the scaffolds fabrication techniques using biocompatible polymers and their biomedical application: A technical and statistical review. Journal of Saudi chemical society. 2020;24(2):186-215.
3. Haider A, Haider S, Kang I-K. A comprehensive review summarizing the effect of electrospinning parameters and potential applications of nanofibers in biomedical and biotechnology. Arabian Journal of Chemistry. 2018;11(8):1165-88.
4. Staudinger H. Über polymerisation. Berichte der deutschen chemischen Gesellschaft (A and B Series). 1920;53(6):1073-85.
5. Haider S, Kamal T, Khan SB, Omer M, Haider A, Khan FU, et al. Natural polymers supported copper nanoparticles for pollutants degradation. Applied Surface Science. 2016;387:1154-61.
6. Haider A, Haider S, Kang I-K, Kumar A, Kummara MR, Kamal T, et al. A novel use of cellulose based filter paper containing silver nanoparticles for its potential application as wound dressing agent. International journal of biological macromolecules. 2018;108:455-61.
7. Saldívar-Guerra E, Vivaldo-Lima E. Introduction to Polymers and Polymer Types. Handbook of Polymer Synthesis, Characterization, and Processing. 2013:1-14.
8. Rao KM, Kumar A, Haider A, Han SS. Polysaccharides based antibacterial polyelectrolyte hydrogels with silver nanoparticles. Materials Letters. 2016;184:189-92.
9. Shokri J, Adibkia K. Application of cellulose and cellulose derivatives in pharmaceutical industries. Cellulose-medical, pharmaceutical and electronic applications: IntechOpen; 2013.
10. Kharkwal H, Malhotra B, Janaswamy S. 1 natural polymers for drug delivery: an introduction. 2017.
11. Olatunji O. Classification of natural polymers. Natural Polymers: Springer; 2016. p. 1-17.
12. Khan MUA, Raza MA, Razak SIA, Abdul Kadir MR, Haider A, Shah SA, et al. Novel functional antimicrobial and biocompatible arabinoxylan/guar gum hydrogel for skin wound dressing applications. Journal of Tissue Engineering and Regenerative Medicine. 2020;14(10):1488-501.
13. Arif U, Haider S, Haider A, Khan N, Alghyamah AA, Jamila N, et al. Biocompatible polymers and their potential biomedical applications: A review. Current pharmaceutical design. 2019;25(34):3608-19.
14. Hocking PJ. The classification, preparation, and utility of degradable polymers. Journal of Macromolecular Science, Part C: Polymer Reviews. 1992;32(1):35-54.
15. Maitz MF. Applications of synthetic polymers in clinical medicine. Biosurface and Biotribology. 2015;1(3):161-76.
16. Parisi OI, Curcio M, Puoci F. Polymer chemistry and synthetic polymers. Advanced Polymers in Medicine: Springer; 2015. p. 1-31.
17. Kamaly N, Yameen B, Wu J, Farokhzad OC. Degradable Controlled-Release Polymers and Polymeric Nanoparticles: Mechanisms of Controlling Drug Release. Chem Rev. 2016;116(4):2602-63.
18. Barnes CP, Sell SA, Boland ED, Simpson DG, Bowlin GL. Nanofiber technology: designing the next generation of tissue engineering scaffolds. Advanced drug delivery reviews. 2007;59(14):1413-33.
19. Wei G, Ma PX. Partially nanofibrous architecture of 3D tissue engineering scaffolds. Biomaterials. 2009;30(32):6426-34.

20. Vasita R, Katti DS. Nanofibers and their applications in tissue engineering. International Journal of nanomedicine. 2006;1(1):15.

21. Ma PX. Biomimetic materials for tissue engineering. Advanced drug delivery reviews. 2008;60(2):184-98.

22. Ma PX, Zhang R. Synthetic nano-scale fibrous extracellular matrix. Journal of Biomedical Materials Research: An Official Journal of The Society for Biomaterials, The Japanese Society for Biomaterials, and The Australian Society for Biomaterials. 1999;46(1):60-72.

23. Lee T-J, Kang S-W, Bhang SH, Kang JM, Kim B-S. Apatite-coated porous poly (lactic-co-glycolic acid) microspheres as an injectable bone substitute. Journal of Biomaterials Science, Polymer Edition. 2010;21(5):635-45.

24. Liu X, Jin X, Ma PX. Nanofibrous hollow microspheres self-assembled from star-shaped polymers as injectable cell carriers for knee repair. Nature materials. 2011;10(5):398-406.

25. Guo B, Ma PX. Synthetic biodegradable functional polymers for tissue engineering: a brief review. Science China Chemistry. 2014;57(4):490-500.

26. Holzwarth JM, Ma PX. Biomimetic nanofibrous scaffolds for bone tissue engineering. Biomaterials. 2011;32(36):9622-9.

27. Zhang R, Ma PX. Poly (α-hydroxyl acids)/hydroxyapatite porous composites for bone-tissue engineering. I. Preparation and morphology. Journal of Biomedical Materials Research: An Official Journal of The Society for Biomaterials, The Japanese Society for Biomaterials, and The Australian Society for Biomaterials. 1999;44(4):446-55.

28. Boccaccini AR, Blaker JJ. Bioactive composite materials for tissue engineering scaffolds. Expert review of medical devices. 2005;2(3):303-17.

29. Lei B, Shin K-H, Noh D-Y, Jo I-H, Koh Y-H, Choi W-Y, et al. Nanofibrous gelatin–silica hybrid scaffolds mimicking the native extracellular matrix (ECM) using thermally induced phase separation. Journal of Materials Chemistry. 2012;22(28):14133-40.

30. Naeem M, Bae J, Oshi MA, Kim M-S, Moon HR, Lee BL, et al. Colon-targeted delivery of cyclosporine A using dual-functional Eudragit® FS30D/PLGA nanoparticles ameliorates murine experimental colitis. International journal of nanomedicine. 2018;13:1225.

31. Naeem M, Lee J, Oshi MA, Cao J, Hlaing SP, Im E, et al. Colitis-targeted hybrid nanoparticles-in-microparticles system for the treatment of ulcerative colitis. Acta Biomaterialia. 2020;116:368-82.

32. Naeem M, Awan UA, Subhan F, Cao J, Hlaing SP, Lee J, et al. Advances in colon-targeted nano-drug delivery systems: Challenges and solutions. Archives of pharmacal research. 2020;43(1):153-69.

33. Haider A, Versace D-l, Gupta KC, Kang I-K. Pamidronic acid-grafted nHA/PLGA hybrid nanofiber scaffolds suppress osteoclastic cell viability and enhance osteoblastic cell activity. Journal of Materials Chemistry B. 2016;4(47):7596-604.

34. Haider A, Kim S, Huh M-W, Kang I-K. BMP-2 grafted nHA/PLGA hybrid nanofiber scaffold stimulates osteoblastic cells growth. BioMed research international. 2015;2015.

35. Haider A, Gupta KC, Kang I-K. PLGA/nHA hybrid nanofiber scaffold as a nanocargo carrier of insulin for accelerating bone tissue regeneration. Nanoscale research letters. 2014;9(1):1-12.

Chapter 16
Biotechnology-Based Therapies

Behzad Qureshi, Saadullah Khan, Zia ur Rehman, and Noor Muhammad

Abstract Pharmaceutical biotechnology is an emerging field that applies biotechnological principles for the development of drugs. There are many established biopharmaceutical products developed through recombinant DNA technology over the course of decades. This first section of the chapter depicts a basic knowledge about molecular biology and biotechnological techniques. Later the chapter discusses recombinant DNA technology and its applications in therapeutics and drug production. Finally, some of the well-known examples of biotechnology-based products are also explained in the chapter.

Keywords Recombinant DNA technology · Gene therapy · Oligonucleotides · Antisense/ribozymes · Aptamers · Hematopoietic growth factors · Cytokines · Monoclonal antibodies · Vaccines

16.1 Introduction

Biotechnology has been defined in many different ways since the inception of the field. It can be defined as "the technology that utilizes cellular and biomolecular processes for development of products that has applications in medicine, food and energy." Pharmaceutical biotechnology is an emerging field that applies biotechnology for development of drugs [1].

Civilizations as early as 8000–500 BCE used biotechnology directly or indirectly when they domesticated crops and livestock, fermented bear using yeast, produced cheese, fermented wine, selectively bred of plants through pollination, and treated boils using moldy soybean curds (GTB). Ever since the discovery of

B. Qureshi · S. Khan · Z. u. Rehman · N. Muhammad (✉)
Department of Biotechnology and Genetic Engineering, Kohat University of Science and Technology, Kohat, Pakistan

© The Author(s), under exclusive license to Springer Nature Switzerland AG 2022
S. A. Khan (ed.), *Essentials of Industrial Pharmacy*, AAPS Advances in the Pharmaceutical Sciences Series 46, https://doi.org/10.1007/978-3-030-84977-1_16

261

biochemical conversion with enzymes by Louis Pasteur (in 1870), biotechnology has emerged as a new field of science. However, the modern era of biotechnology started after the discovery of DNA (in the late 1860s), followed by structure elucidation of DNA by Watson and Crick (in 1953).

Understanding the structure of DNA and the molecular process of the cell brought the notion that DNA encodes proteins and, in this way, controls all cellular processes. During the 1970s, the world was introduced to recombinant DNA technology which allowed biotechnologists to control and manipulate gene expression in organisms which was used for biological applications [2]. Afterward, through constant intense research around the world, many biotechnological developments empowered us to manipulate cellular and bimolecular process with precision and accuracy.

Prokaryotic cells are unicellular organism that lacks membrane-bounded internal structures such as nucleus and other organelles (Fig. 16.1). These cells are small in size (around 0.1–5 µm) and are much simpler mechanistically. These cells lack specialized organelles, such as mitochondria (energy factory), and instead use cellular membrane for production of ATP (energy packet) [(3]). Prokaryotic cells have a small size genome, quick reproduction cycle, and quick growth rate compared to eukaryotic cells. *Bacteria* and *Archaea* are the most common and famous examples of prokaryotes.

The Eukaryotic Cells are upgraded and complex version of prokaryotic cells. These cells have a defined nucleus and other organelles enclosed within a plasma membrane. All the organelles have a specific structure and functions. These functions include, among others, energy production, waste removal, protein synthesis, and protection against pathogens.

Eukaryotic cells have a bigger genome size (10–1000 µm), slow growth, and slow reproduction as compared to the prokaryotic cells. Cells of animals, plants, fungi, and protists are composed of eukaryotic cells.

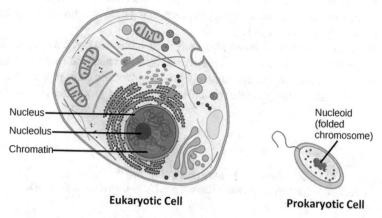

Fig. 16.1 Basic structure of eukaryotic cell and prokaryotic cell

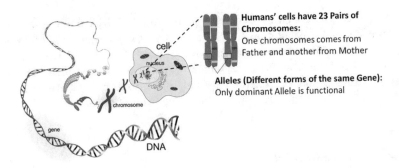

Fig. 16.2 Genomic components of the eukaryotic cell

DNA is a ladder-shaped molecule with a helical orientation, composed of complementary long chains of nucleotides (basic building blocks of nucleic acid). Nucleotides are linked by phosphodiester bond with each other and are attached by specific hydrogen bonding with complementary strand. DNA holds genetic information that is passed on to the next generation through cell division. All the cellular structures and functions are governed by DNA, in the form of codons (combination of three nucleotides) [4]. These instructions are expressed to RNA (ribonucleic acid, a cousin molecule of DNA) through the processes of transcription and subsequently translated into protein, the ultimate end product of gene expression. These series of events inside a cell are known as central dogma of molecular biology (Fig. 16.2).

The basic constructive unit of DNA is *nucleotide* that is made up of a phosphate group, a sugar group, and a nitrogen base. There are four types of nitrogen bases in DNA: adenine (A), thymine (T), guanine (G), and cytosine (C). These nucleotides are bonded together in double-stranded helical fashion by specific hydrogen bonds between the basis. Adenine is complementary to thymine while guanine is complementary to cytosine [5]. However, in ribonucleic acid (RNA) thymine is replaced with uracil. The combination of these nucleotides is responsible for all features and characteristics of organisms. The specific combination of nucleotides that express a specific trait is called *gene* [6]. Human DNA has around 3.2 billion base pairs. Every feature of the human, for instance, the eye color, height, skin tone, or the facial features, is a result of a specified gene or combination of genes.

16.2 DNA Replication

All the structural and functional information of cells are stored in its DNA. This information is transferred from parents to offsprings through the process of cell division. The information is expressed in cells by the process of transcription and translation.

First, an enzyme *helicase* unwinds the double-stranded DNA, thus creating space between the two antiparallel strands. Subsequently, *DNA polymerase* acts by adding deoxyribonucleotide monomers to the growing strand of DNA at 5′ end. DNA replication starts from a specific site called *origin of replication (ori)* [2]. Small genomes, such as that of bacteria, have only one ori site present, whereas larger genomes, such as humans' genome, can have hundreds of ori sites.

16.2.1 Transcription

Transcriptions starts with binding of enzyme *RNA polymerase* on specific sequence of nucleotides at a specific site of gene, called *promoter region*, the regulatory part of the gene [7]. Based on RNA polymerase binding efficiency and multiple other factors the promoters can be either, strong promoters and weak promoters which, respectively, give more products or less products.

Cells make a complementary RNA, i.e., *messenger RNA* (mRNA), of a specific segment of DNA (the gene), as shown in Fig. 16.3.

RNA polymerase binds to the double helix of DNA; it partially unwinds and transcription process starts subsequently. The building blocks of RNA synthesis are ribonucleotides ATP, GTP, CTP, and UTP. The developing RNA strand is complementary to DNA template strand and is in 5′ to 3′ direction. This means that G in the DNA strand will form C on the RNA. Similarly, C and T on DNA strand will add G and A in RNA strand; however, A on DNA yields U on RNA strand. The transition stops due to an intrinsic feature of the gene, either presence of stop codon or

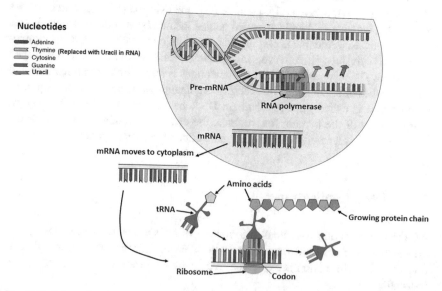

Fig. 16.3 Schematic representation of transcription and translation process

external intervention of specific terminating proteins [8]. The transcription process occurs inside the nucleus of the cell. Transcription is highly exploitable in biotechnology.

RNA contains a triplet combination of nucleotides, called the *codon*, that codes for a specific amino acid. For instance, the codon for methionine is ATG. There are 64 codons which code for 20 amino acids. Besides, RNA also contains *start codon and stop codon*, which are used for ribosomal instructions [9]. This codon language for synthesis of amino acid sequence is universal, i.e., similar in almost all the living organisms.

16.2.2 Translation

The mRNA carries genetic information from DNA to cytoplasm, where proteins are made according to the information stored in mRNA sequence. This process is called *translation*.

mRNA attaches itself to ribosome and passes through the ribosome. Another type of RNA is tRNA, which carries a specific amino acid and binds to a matching sequence of mRNA. Every tRNA has a unique triplet loop known as *anticodon*. When the complementary anticodon of tRNA and codon of the mRNA coincides, the amino acid is unloaded from tRNA and is attached to the polypeptide chain of protein. This cycle continues until the *stop codon* arrives, and the protein is released for further processing [10].

Translation is a highly coordinated complex cellular process that involves mRNA molecules, ribosomes, tRNA molecules, amino acids, aminoacyl synthetases, and a number of translation factors. The translation process happens outside the nucleus, in ribosome, which is why ribosome is considered as assembly hub for all the processes involved in protein synthesis.

16.3 Recombinant DNA Technology

Genetic manipulation in any cell cannot be done by just injecting DNA into cells, since DNA injected into a cell without the aid of certain factor will be immediately degraded by cellular nucleases. Scientist discovered methods for synthesis of DNA molecule by genetic recombination of the DNA molecules from two different species; the resultant DNA is called *recombinant DNA (rDNA),* and the technology used for this DNA manipulation is known as *recombinant DNA technology* [11].

Successful transfer of DNA to a cell or organisms requires a carrier.

Replicons used as carriers for foreign DNA fragments are termed *vectors*, which are mainly comprised of plasmids from bacteria, yeast, bacterioviruses, animal viruses, or plant viruses.

DNA of interest is isolated and is stored at cold temperatures to avoid degradation. Then an appropriate plasmid is isolated which can be successfully fused with the DNA fragment of interest.

Endonuclease or *restriction enzymes* (DNA scissors) and ligase (DNA glue) are two main enzymes involved in the synthesis of rDNA. Using restriction enzyme DNA can be cut into two or more fragments. Similarly, plasmid is also cleaved at a very specific site. The desired DNA fragment and cleaved plasmid are joined together by the action of ligase. *Ligase* is a class of enzyme that joins the phosphodiester bond of cleaved DNA avoiding any gaps or nicks.

16.3.1 DNA Transfer

For rDNA to work, it has to be successfully transferred to a host cell for expression.

Some species of bacteria, such as *B. subtilis*, are able to take up rDNA under physiological conditions. This form of rDNA uptake is known as *natural transformation*. However, *B. subtilis* is not always an appropriate candidate because of the nature of the desired output; hence, in many cases bacterial cells of different species are forced for the process of rDNA uptake inside the cell. One way of increasing the rDNA uptake is called *heat-shock method* [12]. In this method, a heat shock is provided to the bacterial cell in the presence of Ca^{2+} which facilitates the DNA uptake process. Another way of increasing rDNA uptake in a bacterial cell is by *electroporation*, i.e., the mixture of rDNA and cells is exposed to an electrical current. The electrical current partially create pores in the cell membrane, which favors DNA uptake by the cell. Another more natural method of DNA transfer is by inserting rDNA into a *bacteriophage* that infects the bacteria and eventually integrates rDNA into the bacterial genome.

Conjugation is a process in which there is a direct cell-cell contact accompanied by the exchange of genetic material. This process can be used to transfer rDNA to a desired bacterial cell which requires a special plasmid molecule known as *conjugative plasmid*. Conjugative plasmid is a class of plasmids which can jump from a donor bacterial cell to a recipient bacterial cell during conjugation. In this way, desired gene can also be transferred to bacterial cell using *conjugative plasmid*.

Apart from the above-discussed natural ways of transferring rDNA to a desired cell, mechanical ways, such as *microinjection*, can also be used to transfer DNA directly into the nucleus, using microneedles.

16.3.2 DNA Sources

For synthesis of rDNA, a desired segment of DNA (gene) is required. Isolating desired gene from a large genome is a tedious process, and successfully isolating the desired gene is very challenging. The common sources for gene isolation are bacteria, plant cells, or animal cells. However, researchers have developed techniques to simplify the process.

Synthetic DNA: Gene can be synthesized directly from a known DNA sequence or indirectly from known amino acid sequence. This DNA synthesis technique typically works for small protein sequences [13].

cDNA: As discussed previously, DNA transcribes into mRNA that translates into proteins. This mRNA if isolated from a cell can be converted back into DNA molecule with the help of enzyme reverse transcriptase. Briefly, mRNA isolated from the cell is treated with reverse transcriptase, which synthesizes a complementary DNA strand attached to RNA molecule. The RNA part is removed by treating the RNA-DNA hybrid with alkali. A primer is attached to a single-stranded complementary DNA, and DNA polymerase is allowed to synthesize the second complementary strand, thus yielding a double-stranded complementary DNA (cDNA). A promoter sequence is attached at the start of cDNA, since the gene cannot be expressed without promoter region.

DNA Libraries: A DNA library is a collection of DNA segments of the total DNA of an organism that have been cloned into vectors so that the desired DNA fragments can be isolated for further research. These libraries are either genomic libraries or cDNA libraries [14]. Genomic libraries contain large fragments of DNA incorporated into vectors, while cDNA libraries are made from cDNA cloned into plasmid vector. cDNA libraries contain much smaller fragments than genomic DNA libraries, since the former lack non-expressed genomic regions, such as introns, and 5′ and 3′ noncoding regions.

16.3.3 Production by Recombinant DNA Technology

The possibility of any protein as a potential drug and its subsequent clinical use primarily depends on the availability of sufficient quantities of the particular protein. This is best achieved by recombinant means.

Production of human insulin in microbial cells was the first commercial application of recombinant DNA technology. The main role of insulin is to regulate glucose level in blood. Before the introduction of recombinant DNA technology, the insulin for treatment of insulin-dependent diabetes mellitus was obtained by extraction from porcine or bovine pancreas. However, most of the approved protein-based drugs including insulin are produced in engineered cell lines by recombinant means.

Human insulin is a 51-amino-acid dimer, containing two chains, chain A and chain B comprising 21 and 30 amino acids, respectively. Both of these chains are held together by S-bond between the amino acid cysteine present on the opposite chains.

The strategy to synthetically produce insulin using recombinant DNA technology is to produce chain A and B separately as observed naturally. The chain A part of DNA was linked together using a specific set of oligonucleotides. The synthetic DNA fragment was then integrated to the end of lacZ gene using ligation fusion.

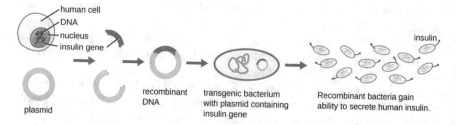

Fig. 16.4 Basic representation of recombinant DNA technology used for Insulin production

This lacZ gene is controlled by lac promoter (Fig. 16.4). This whole fragment of DNA was inserted and fused with pBR322 cloning vector of *E. coli* [15]. Then, a codon for amino acid methionine was inserted at the fusion point of lacZ and chain A, for a smart reason which will be explained in the lines ahead. Afterward, same technique was performed for the production of B chain. The DNA fragments for the N-terminal and C-terminal were independently synthesized and then fused together and eventually integrated in a separate molecule of pBR322 plasmid. Similar to the case of chain A, a methionine codon was also added here between lacZ and the DNA fragment of chain B.

LacZ is the part of well-known lactose operon. There are two advantages of using this operon. Primary advantage is that both DNA fragments depend on lac promoter and lacZ gene for their expression, which allows effective and controlled expression. Secondary advantage is lacZ codes for β-galactosidase which helps protect chain A and B from proteolytic breakdown.

These plasmids are then cloned in bacteria and subsequently expressed. After expression, the chain A and chain B are fused with the product of lacZ which is β-galactosidase. Since chain A and chain B are attached to β-galactosidase it means protein cannot work properly unless the β-galactosidase part is removed. Here comes the interesting catch, to separate chain A and B from β-galactosidase, an inorganic compound cyanogen bromide. This agent has the ability to cleave a peptide when it encounters amino acid methionine. The smart move here is that neither of the chains A or B has a methionine amino acid so the agent does not interact with structure of the rest of the protein. The last step is to mix A and B chains allowing them to form S bonds with cysteine amino acids spontaneously.

16.4 Specific DNA Techniques

16.4.1 PCR Technology

Kary Mullis, an American biochemist, invented polymerase chain reaction (PCR) to make many copies of specific segment of DNA in vitro. He received the Nobel Prize in 1993 for his invention.

PCR technique involves repeated thermal cycling of desired DNA [16]. Briefly, isolated DNA is mixed with primers dNTPs, Taq DNA polymerase, and $MgCl_2$ in a reaction tube which contains nuclease-free water. This PCR tube is then placed in a thermal cycler device. The underlying process involves three crucial steps.

Thermal Denaturation Firstly, the desired double-stranded DNA is separated into single strands by exposing the DNA to 92 °C.

Primer Annealing In this step, primer is attached to specific part of the desired DNA, which serves as a starting point for DNA amplification. In this step, the temperature of the thermal cycler is dropped automatically to 55 °C, which is the optimal temperature for primer annealing, however, it can also be changed depending on the access of certain nucleotides in the primer sequence and hence template DNA sequence.

DNA amplification The third and last step involves the extension or amplification of the target DNA. The temperature is adjusted to 72 °C for the DNA amplification.

All of these three steps are repeated for 35 to 40 cycles, the corresponding thermal cycles are pre-adjusted in the thermal cycler. This whole process yields millions of copies of the desired DNA.

It is important to note that PCR technique involves thermal cycling at high temperature. Normal DNA polymerase is degraded at high temperature. Hence, Taq DNA polymerase, a highly heat-resistant DNA polymerase extracted from bacterium *T. aquaticus*, is used in this technique. Since *T. aquaticus* is found in hot springs, thus the isolated Taq DNA polymerase works best at high temperatures, making it suitable for PCR reaction.

16.4.2 *DNA Hybridization*

DNA usually exist as double-stranded molecule. The two strands mutually bind in a complementary fashion by a process called *hybridization*. During DNA replication, hybridization of new strands to the old strands is a natural process. However, this phenomenon can be exploited by preparing nucleic acid probes to confirm the presence or absence of certain gene in cells, for instance, diagnostic test for cystic fibrosis (CF).

16.4.3 *DNA and Genome Sequencing*

DNA sequencing is the process of determining the nucleotide sequence in a segment of DNA.

The classical sequencing technique was introduced by British biochemist Fred Sanger and his colleagues in 1977, known as *Sanger sequencing*. Novel, large-scale approaches have been developed to reduce the cost of DNA sequencing, known as *next-generation sequencing techniques* [17].

In Sanger sequencing DNA segments of 900 base pairs in length are routinely sequenced. In human genome project, Sanger sequencing was used to sequence small fragments of human DNA.

Sanger sequencing uses a method called chain termination which is based on DNA replication enzymology. Sanger sequencing depends on a special form of nucleotides which are called dideoxyribonucleotide triphosphate (ddNTP). These are synthetically produced analog of the original naturally existing dNTP. This analog not only lacks the 2′ hydroxyl group on the ribose but also lacks 3′ hydroxyl group which is why it is called ddNTP. But why ddNTP, why not the original dNTP? The special feature of ddNTP is that DNA polymerase can incorporate it to DNA strand, but once ddNTP is incorporated, DNA chain elongation can no longer take place because DNA polymerase requires a normal dNTP with 2′ hydroxyl group missing. This means there is not a suitable substrate available for the proper activity of DNA polymerase. This chain termination is used to identify the sequence of a DNA molecule. Four tubes are prepared which have template DNA strand, primer, and four different dNTPs (dATP, dGTP, dCTP, and dTTP). Furthermore, ddNTPs are also added to all the four groups in such a fashion that every tube contains a different ddNTP. First tube has ddATP, second tube has ddTTP, third tube has ddGTP, and the fourth tube has ddCTP. The amount of dNTP and ddNTP is adjusted to a level in such a manner that a small number of DNA templates will incorporate the specific ddNTP at a specific point which will lead to chain termination. In each tube, the specific ddNTP will terminate the chain which can then reveal that at a particular point, A, T, C, or G is incorporated since we know already which tube has ddNTP. After the reaction has taken place, the reaction tube content is subjected to run on a high-resolution polyacrylamide gel electrophoresis system which separates DNA strands on the their respective sizes. Tube 1 reveals the positions of A, tube 2 of T, tube 3 of G, and tube 4 of C. Every tube is separated individually such that the content of all the tubes is not mixed which will cause confusion. The separated DNA strand in polyacrylamide gel electrophoresis can be visualized by autoradiography which has radioactively labeled dNTP or by fluorography in which a fluorescent group is chemically attached to the primer. In this manner, one can know the sequence of a DNA fragment of the size ranging from a few hundred base pairs to a few thousand.

Once scientists learned how to know the nucleotide sequence of a DNA fragment of a specific length, the next obvious goal was to know the whole genome of an organism. Genome is the total genetic material present inside a cell or organism. Compared to a fragment of DNA, genomes are much larger in size. A new and fast method was needed to sequence whole genomes. Next-generation sequencing uses high-throughput technologies to sequence whole genomes. This technology enabled and made possible the sequencing of whole genome reliably and with fast pace. This technology uses complex molecular biology and sophisticated machines combined with knowledge of computational biology; it revolutionized the sequencing method and paved new ways in pharmaceutical biotechnology. Such a fast and sophisticated technologies boosted the pace of fields of genomics and proteomics to understand the inside unseen complex processes of organism. Whole genome sequencing enables researchers to understand any organism at the very genetic and cellular level. This genetic and cellular knowledge is vital due to the fact that it can enable us to design and discover new drugs for different diseases. For example, traditional antibiotics are secondary metabolites that are isolated through a laborious downstream process. These antibiotics act on vital process in an organism which causes the death of the pathogen. But these are limited to act on a few target molecules of the pathogens. The problem is once the pathogen develops a certain resistance to the antibiotic, we do not have any other options to try since the traditional antibiotics are the only option. Understanding the genomics and proteomics of an organism, let us develop new synthetic drugs based on the knowledge about the new target molecules that were identified through the sequencing technology.

16.4.4 Cell Cultures

Cell culturing is an important component of biotechnology products, since biological drugs are primarily produced within the prokaryotic cells, eukaryotic cells, microbial cells, plant cells, or animal cells. Different types of cells require different culturing conditions.

16.4.4.1 Microbial *Cell* Culture

There are numerous microbes discovered, but only a few are favored for biotechnological process. Traditionally used microbial cells include but are not limited to *Clostridium acetobutyricum*, *Corynebacterium* sp., *Xanthomonas* sp., *Bacillus* sp., *Lactobacillus* sp., and fungi such as *S. cerevisiae* (baker's yeast), *Penicillium* sp., *Aspergillus* sp. etc.

Bacteria typically grow in four stages; they do not start growing instantly after inoculation in the medium. Initially, they adjust to the media and the culture conditions; this resting stage is known as *lag phase*. The second stage is the *exponential phase*, where bacterial cells grow exponentially, followed by the stationary phase where no increase in number of bacteria occurs. The stationary phase is the most critical phase, since microbes usually release secondary metabolites during this phase that are of great importance in pharmaceutical biotechnology. The last stage is the death phase where cell number decreases due to the total media depletion. These four phases of bacterial growth are jointly known as the bacterial *growth curve*. This curve is highly dependent on the pH, temperature, medium type and concentration, and aeration of the tank.

The growth of microbes is very rapid and they do not require hectic culture procedure and, thus, are considered favorable candidates for isolation of biotechnological products.

16.4.4.2 Animal Cell Culture

Animal cells are complex in nature compared to single-celled bacteria. Hence, the culturing process is also complicated. The cells are isolated from tissue by treating with protease. These isolated cells are grown in specific liquid or solid growth medium, regarded as a *primary cell culture*. These cultures are very delicate and do not survive for longer time, which makes them less promising for biotechnology.

Cells like 3T3 fibroblasts which are nonmalignant immortalized cells are of much interest because of their long-lasting surviving ability when provided with specific growth medium and conditions.

Animal cell culture requires specific liquid or solid media and culturing conditions. Moreover, optimum pH of 7.0 and isotonic conditions are also needed.

16.4.4.3 Cell Culture

Prehistorically and till date, plants are an indispensable source of pharmacologically active compounds used in different treatments. These compounds are often complex in nature which makes it difficult to be chemically synthesized particularly on commercial scale. Even if extracted, these active compounds are low in quantity, and this very reason triggered researchers to explore alternatives for large-scale production.

Since plants cells are as complex as animal cells, it is highly difficult to grow them in large number and maintain these cells for production since researchers have tried it. Apart from this problem, it is really hard to maintain plant cells in a specific differentiated state. Usually, a specific compound is produced at a specific differentiated state which when lost, the cells then do not produce that very compound which makes the cells redundant. Since plant cell culture requires hormones like auxins and cytokinins, this might be an explanation of the differentiation state problem. Due to these complexities, efficient production of pharmaceutically interesting compounds in plant cell culture system is rare. To encounter these issues, a system of repeatedly selecting cell lines with the highest production potential is subcultured and subsequently used.

A better understating of cellular differentiate combined with genetic understandings needs to be tailored together to have a large-scale view and then subsequently exploited to overcome current culturing and production issues.

16.5 Protein Engineering

Protein engineering is the alteration of specific sites of a gene sequence encoding the transcription to a protein, and these alterations are controlled and calculated to improve the performance and effectiveness of a protein. This strategy is dependent on computational biology tools and wet lab techniques. The approach here is to change amino acids of the protein at calculated site by not losing the actual function of the protein but only to enhance it. This is done through changing the DNA sequence of the gene precisely that codes for the protein of interest, to obtain the engineered protein as designed, since we already know that proteins are transcribed from DNA sequencing in the form of MRNA and then translated to proteins using cellular translational machinery. In the sector of drug development, protein engineering is successfully used to develop recombinant proteins, ultimately leading to improved pharmacodynamics and pharmacokinetic profiles. Improvement in pharmacodynamic profile can help us in obtaining a drug with faster or slower action, depending on the nature of treatment it is used for, the alteration of pharmacological half-life and kinetics. Furthermore, the engineering of protein is used for receptor binding specificity, reducing immunogenicity, and enhancing physical and chemical protein shelf half-life.

Muteins are those proteins that are engineered through site-directed mutagenesis. Site-directed mutagenesis is the changing of DNA sequencing at specific sites which can ultimately code for the protein with specific engineered mutations. Let us take in account the example of insulin. Native insulin associates from dimers up to hexamers at high local concentrations are what are usually found at the site of injection, leading to retarded dissolution and activity in the body. As a result of structure elucidation, proline and lysine at positions 28 and 29, respectively, in the B chain were identified to play a crucial role and were therefore subjected to site-directed mutagenesis. Switching B28 and B29 of proline and lysine reduced the association affinity 300-fold, resulting in faster uptake and action, as well as shorter half-life.

Protein engineering used to be a difficult approach, but with recent advancement in technology, it is now achievable.

16.5.1 Post-translational Engineering

When an mRNA is translated into a protein through translation process, the protein is subsequently exposed to post-translational modification within the cell and it is a natural process. Post-translational modification is the covalent attachment of a chemical group. These chemical groups include phosphate group, glycans, lipids including fatty acids and cholesterol, etc. This natural modification of protein can help achieve different functions and can change the behavior of a protein.

Post-translational engineering is the artificial modification after the biosynthesis of a protein. It involves attachment of chemical groups to a protein as mentioned earlier or modifying a preexisting chemical group.

Glycosylation is the most complex and extensively occurring post-translation modification naturally. High frequency of occurrence makes glycoengineering of greater interest. Alternation in the glycosylation pattern at the protein backbone can remarkably change the pharmacokinetic profile. Approximately 40% or more approve proteins are glycoengineered. Mammalian cells like "Chinese hamster ovary cells or baby hamster kidney cells" are used as factory for the production of these post-translation engineered proteins.

Another post-translation engineering chemical group that is frequently used is attaching polyethylene glycol (PEG) known as PEGylation. PEGylated proteins are promising specially in pharmaceutical biotechnology due to improved protein solubility, improved thermal and mechanical stability, and reduced immunogenicity with many more qualities.

Novo Nordisk's Victoza® (liraglutid) is a non-insulin once-daily medication which can help improve sugar levels in adults with type II diabetes. This medication carries glucagon-like peptide 1 (GLP-1) analog having a 97% protein sequence homology. It has an attached C16 fatty acid at Lys26.

16.5.2 Synthetic Biology

Synthetic biology is a cross-disciplinary area of research which involves creating new or advance biological part devices and systems and to redesign and reengineer existing natural biological systems. It is a collaboration between biotechnology, genetic engineering, molecular biology, biophysics, systems biology, computational biology, electrical and computer engineering, and evolutionary biology. This is an advance emerging area of research. Synthetic biology allows standardization of biological devices known as "biobricks." These biobricks include promoters, coding DNA sequence, transcription factors, terminators, ribosomal binding sites, plasmids, primers, and translational units. This synthetic technology massively relays on DNA sequencing, protein engineering, and assembling the designed biobricks to fabricate a production host of interest.

Synthetic biology is becoming more and more interesting from a standpoint of pharmaceutical biotechnology. There are many examples of pharmaceutical biotechnology products produced through synthetic biology and one of them is artemisinic acid. This molecule is precursor to the biosynthesis of artemisinin, which is an antimalarial drug, and this artemisinic acid was successfully transferred into *E. coli* for bio-production in 2004. Synthetic biology is a promising avenue when it comes to designing and producing new drugs.

16.6 Gene Therapy

Gene therapy is the direct involvement of nucleic acids as therapeutic agents. There are some genetic diseases that are not curable completely through traditional medication. Although symptomatic treatment is possible through conventional drugs, it is not a permanent cure. Gene therapy offers a unique approach to fix or replace the abnormal/mutated part of the DNA. Since, when there is a genetic disease, it usually involves a faulty or missing protein in a particular tissue or system. Using gene therapy, one can introduced a normal copy of the faulty gene to the cells to restore the normal function of the protein [18]. Direct injection of the genetic material is not always prolific when it comes to gene therapy. Hence, we use vectors for a successful transfer and integration of DNA into the cell. Enough information must be acquired about the gene, the cell type being treated, and their environment. Once the gene is identified, a normal copy of this gene is cloned into the vector. Frequently used vectors are modified, non-harming viruses. These viral vectors are precisely modified and tailored to the cell type to have a targeted approach. Once the vector is ready, it is injected directly into the tissue with abnormal cells. Virus infects the cell by injecting the genetic material into the cell and integrating in the chromosome. After integration into the chromosome, the gene expresses itself and starts producing a normal functioning protein. It is important to note that some adenovirus introduces their genetic material in the nucleus rather than chromosomes, and these viruses are also used for gene therapy according to the requirement.

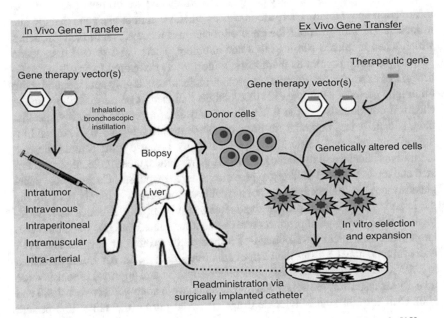

Fig. 16.5 Illustration of in vitro and in vivo gene therapy, adopted from Daan J.A et al., [18]

Again, the vectors can be injected intravenously (by IV) in the affected tissue where it can target the cells in the tissue (in vivo approach), or alternatively, genetically abnormal cells are harvested from the patient's tissue/body and then exposed to the vectors which can then affect these cells by injecting the genetic material (in vitro approach), as shown in Fig. 16.5. These cells are tested in the laboratory settings, and if the cells have the new genetic material and are producing a normal protein, then they are transferred back to the patient where they can compensate for the missing or faulty protein.

Gene therapy is being used and tested for many of the genetic diseases, and more than 60% are being tested for cancer treatment, since cancer cells arise from mutations in the genome of cells.

16.7 Protein Therapy

Ultimately, any process going on in the body is associated with proteins directly or indirectly. Particularly, whenever we have a faulty gene in a cell or group of cells in a tissue, this indicates that there is some protein whose structure is disturbed; hence it is not functioning properly. An estimated 20,000 to 25,000 genes are present in *H. sapiens* which codes for estimated 80,000 to 400,000 proteins through alternative splicing and post-translational modification. Clearly, without proteins, vital cell or body functions cannot be executed at all. Protein therapy involves oral intake or injection of therapeutic and recombinant proteins to a patient with faulty proteins majorly due to a genetic defect. But why not gene therapy instead of protein therapy? Cell and genome are very complex structures and we do not have complete information about them. There are instances when we cannot use gene therapy due to lack of knowledge about the gene or sometimes there are multifactorial genes which when tempered can lead to a new problem. There exist several therapeutic advantages using protein as therapeutic molecules as compared to small-molecule drugs. For example, the function of protein molecules is highly specific, and more often they perform a complex set of functions that cannot be mimicked by simple chemical compounds. As the function or catalysis of a protein is highly specific, there is usually less potential for protein therapeutics to interfere with normal biological processes and cause adverse effects. Also, because protein molecules are naturally produced and the same are used as therapeutics, these molecules are tolerated and are considered unlikely to provoke immune responses. Currently for many genetic disorders, gene therapy is not available or considered as highly risky for the participating individuals; protein therapeutics can be highly effective as replacement of the missing gene products due to mutation in the gene (s). We should also know that the clinical development of a therapeutic protein or enzyme and FDA approval time may be faster than that of small-molecule drugs. An article published in 2003 revealed that the average clinical development and approval time was more than 1 year faster for 33 protein therapeutics that were approved between 1980 and 2002 than for 294 small-molecule drugs approved during the same time period.

Similarly, as we know that proteins are unique in form and specific in their function, companies and industries can obtain long-lasting patent protection for protein therapeutics. Fast track approval from the FDA and far-reaching patent protection of protein therapeutics make proteins attractive from a financial perspective compared with small-molecule drugs.

16.7.1 Oligonucleotides

Short-lengthened chains of chemically modified ribo- or deoxyribonucleotides are called "oligonucleotides" or simply "ONs." Their length is usually between 15 and 25 bp. ONs can bind to chromosomal DNA or mRNA through Watson-Crick and Hoogsteen base pairing. This pairing of ONs to DNA or mRNA opens opportunities for highly specific interventions in gene transcription, translation, repair, and recombination for therapeutic applications. Theoretically, a nucleotide sequence of 15 to 25 bp only occurs once in human genome which offers high specificity to target any gene or any portion of a DNA since human genome is 3.2 billion base pairs long. This feature makes ONs highly specific which enables it to be used in therapeutics. But the mode of action of ONs is not straightforward. It can cause unintended binding to partially sequence homological sites which affects the expression of nontargeted genes leading to abnormality in cellular system. Furthermore, ONs are sensitive to nuclease within the cell or in a free system which can lead to their degradation. Over the course of many years, a lot of modifications and optimization have been done to oligonucleotides to overcome these problems.

16.7.2 Antisense/Ribozymes/EGS

ONs have a wide range of practical application both in research and clinical side. One of them is antisense therapy which is an amazing application of ONs. Traditional antisense ONs consist of a 13- to 25-nucleotide fragment of DNA or RNA. This DNA or RNA fragment if complementary to the targeted sense mRNA can hybridize through Watson-Crick base paring. Upon binding to the targeted RNA or DNA sequence, it brings about gene silencing. Over the course of many years, a lot of improvements are made to antisense therapy including locked nucleic acids for better stability, lesser off-target effects, free uptake by the cell, lesser cytotoxicity in nontarget cells, etc.

Based on mechanism of actions, we have three classes of ONs:

I. mRNA-Blocking Oligonucleotides: These molecules physically prevent or inhibit the progression of mRNA splicing or mRNA translation through binding of complementary oligonucleotides with target mRNA sequences.

II. mRNA-Cleaving Oligonucleotides: These molecules will bind (hybridize) with target mRNA and induce degradation of mRNA by recruiting the cytoplasmic nuclease RNase H (Aboul-Fadl, 2005).

III. mRNA-Cleaving Oligonucleotides: These oligos will induce the degradation of mRNA by recruiting nuclear RNase P or by nuclease activity of the nucleic acid itself (ribozymes/DNAzymes).

Antisense therapy has been used to target lncRNA MALATI, which has shown encouraging results indicating that this can be used as an effective anticancer drug. Majority of clinical studies are focused on RNase H activity on antisense ONs. Downregulation of proteins and mRNA expressions through RNase H-mediated knockdown generally reaches >80%. Most inhibiting ONs are targeted at AUG initiation codon region or within the 5′-untranslated region because it has been reported that apparently ribosome is able to remove hybridized antisense ONs in the coding region. It is worth mentioning that we have approved antisense drug in the market for the treatment of cytomegalovirus-induced retinitis in AIDS patients: fomivirsen (Vitravene) [19].

This is currently one of the promising technologies that can open new treatment ways and allow to treat fatal conditions like cancers.

16.7.3 Aptamers

Aptamers are single-stranded oligonucleotides composed of DNA or RNA folded in a well-defined three-dimensional shape and structure. Generally, DNA- or RNA-based aptamers are 60 nucleotides in length. Based on this length, there are 11,703,240 possible combinations based on the sequence of A-T-G-C, which also implies that we can potentially have 11,703,240 different shapes. This shape/structural diversity enables aptamers to be highly specific. Their binding mechanism to the target molecules depends on the three-dimensional shape, charge, hydrophobic interactions, and hydrogen bridges. Aptamers are artificially isolated, but there are also natural aptamers in riboswitches.

Systematic evolution of ligands by exponential enrichment, or SELEX, is a technique to synthetically generate new aptamers for a specific target. The ligands that show specificity to the target protein are called aptamers. The SELEX procedure is initiated by generating a large library of RNA sequence which is randomized. The size of the library is usually 1015 nucleic acid molecules which are different in their sequences and also folds into different structures based on their specific sequence. All of these different RNAs are tested with their binding affinity with the specific target protein. Those that bind to the protein are separated for those that do not. The binding RNAs are then again subjected to separation for enrichment and to get the best binding RNA molecules. The obtained RNAs are then amplified using reverse transcriptase-PCR technique to get more and more copies of the required RNAs. These resulting cDNA amplicons are in vitro transcribed and generate large quantities of RNAs that bind to the target of interest. The selection process is usually repeated 8–12 times under increasingly stringent binding conditions to achieve a high affinity binding using Darwinian mechanism of selection. This is done to the point when RNA aptamer has the highest binding affinity to the target protein. SELEX can also be done for DNA with the exception of reverse transcriptase-PCR and in vitro transcription process. Today, there are automated systems available to generate new aptamers making the whole process very fast, shifting the required time from months to days. This fast pace generation of aptamers makes this technology a high-throughput technique suitable for creating different aptamers for different targets simultaneously. One problem that has been encountered when creating therapeutic aptamers is that it can be so specific to the human variant of the protein resulting in poor binding affinity within the model organism which is used to test and evaluate the efficacy of the therapeutics. Pegaptanib (Macugen) is an approved aptamer targeting VEGF in wet age-related macular degeneration.

16.8 Biological Drugs

16.8.1 Hematopoietic Growth Factors

The fluid in human veins and arteries is comprised of two major components: i) blood and ii) blood plasma. These components are produced in organs and tissues such as the bone marrow, liver, and spleen. This production system is called hematopoietic system, and the process through which the body manufactures blood cells is known as hematopoiesis. The origin of different cells in blood and plasma is pluripotent stem cells. These cells are found in the bone marrow. All blood cells are divided into three lineages: i) red blood cells (RBCs) which are also known as erythrocytes, ii) lymphocytes which are the cornerstone of adaptive immunity, and iii) myeloid cell lineage which includes granulocytes, megakaryocytes, and macrophages.

These are several conditions, side effects of medicines, and disorders where blood cells function is disturbed. Myelosuppressive chemotherapy, bone marrow transplantation, severe chronic neutropenia, leukemia, and AIDS can cause neutrophil disorders particularly neutropenia. Neutropenia is a serious side effect of chemotherapy. Patients who have neutropenia are not able to fight infections effectively, and if not treated or managed properly, it can be fatal. Through the use of recombinant DNA technology, we now have recombinant human hematopoietic growth factors which are approved in many countries for the treatment and management of several blood-related disorders and conditions. Recombinant human hematopoietic growth factors are identified as "rHu." Recombinant myeloid growth factors rHuG-F and rHuGm-CSF have been tested and used for the treatment of many neutrophil disorders particularly neutropenia.

16.8.2   Cytokines

Isaacs and Lindenmann in 1957 observed a substance that was produced by cell cultures which were infected with virus. This substance interfered with virus-induced infection and thus was called interferon (IFN). In the coming decades, scientists discovered that these substances were actually a group of proteins produced and released by the host cells in response to several viral infections. Interferons interfere in the viral replication resulting in protection from viral infections. IFNs belong to the large class of proteins known as cytokines. These cytokines are basically immuno-communicative substances used by cells to alert the cells in vicinity about a viral infection and provoke a viral defense mechanism within the cells.

16.8.2.1   Interferons (IFNs)

Interferons are proteins produced by eukaryotic cells in response to viral infections, tumors, and other biological inducers. These molecules can promote antiviral activities in neighboring cells and help in regulation of the immune response. These are involved in a variety of other activities and represent a wide family of proteins.

16.8.2.2   Interleukins (ILs)

This is a group of cytokines and are mainly secreted by leukocytes, primarily affecting growth and differentiation of hematopoietic stem cells and immune system cells. These are also produced by other normal cells and cancerous cells. These molecules are of major importance in the regulation of hematopoiesis, inflammation, immunity, embryonic development, and tissue remodeling.

16.8.2.3   Growth Factors (GFs)

These are proteins that activate cellular proliferation and differentiation. Many growth factors stimulate cellular division in several different cell types; others are specific to a particular cell type. These proteins are involved in promotion and proliferation of connective tissue and glial and smooth muscle cells. These can enhance wound healing and are also involved in promotion, proliferation, and differentiation of red blood cells (erythropoietin).

16.8.3 Monoclonal Antibodies

In 1975, two scientists, Milstein and Kohler, developed a breakthrough technique called murine hybridoma technology which enabled the world to reproduce mono-clonal antibodies with unique target selectivity in almost unlimited quantities. But what are monoclonal antibodies and what is their significance? Monoclonal anti-bodies are synthetic antibodies obtained through cloning of a specific white blood cell in a suitable host. These are very specific in host binding, making them very unique for targeting specific regions/sites. There are four major categories of mono-clonal antibodies: (i) murine, made from mouse protein; (ii) chimeric, a hybrid monoclonal antibody having protein parts of mouse and human; (iii) humanized, monoclonal antibodies having small mouse protein portion attached to human pro-teins; and (iv) human, made up of human proteins.

With the advancement in biotechnology, recombinant antibodies (rAbs) are now being produced using synthetic genes. The interesting thing here is, unlike hybridoma-based monoclonal antibodies, rAbs do not need animals or hybridomas in the production process. The procedure to make rAbs is to first isolate the specific gene from the source cell, amplifying and cloning the genes into an appropriate phage vector. Subsequently, the vectors are introduced into a host usually bacteria, yeast, or mammalian cell lines, allowing the expression of the rAbs in adequate amounts.

16.8.4 Vaccines

The purpose of vaccination is to prevent infectious diseases. Vaccination can be considered as one of the most successful prevention strategies in the field of medical sciences. Traditional vaccines that are routinely applied have proved very effective in preventing several infectious diseases. Global eradication of the smallpox due to mass vaccination in the 1970s is one of the key examples. Other infectious agents, like diphtheria, tetanus, poliomyelitis, measles, mumps, and rubella, are under con-trol in almost all developed countries and several developing countries, due to the mass vaccination programs of childhood vaccines. Recently vaccines are not only developed against infectious diseases, but this includes against drug abuse (nicotine, cocaine) and vaccines against allergies, cancer, and Alzheimer's disease. In the rap-idly evolving field of new vaccine technologies, one can discern the improvement of existing vaccines and the development of vaccines for diseases against which a vac-cine is not available yet. Modern developments in biotechnology have a huge impact on current vaccine development. The elucidation of the molecular structures of pathogens and the tremendous progress made in immunology and understanding of the detailed mechanism of immunological system and processes over the past decades have led to the identification of protective antigens and ways of administer-ing them. Along with technological advances, this has caused a shift from empirical vaccine development to more rational approaches, expressing epitopes of antigens (¼ of the smallest molecular structures recognized by the immune system) and/or isolating those antigens that confer an effective immune response and eliminating the structures that cause deleterious effects. Thus, "cleaner" and well-defined products can be obtained, resulting in improved safety. In addition, modern methodologies can provide simpler production processes for selected vaccine components. It is not our intention to provide a complete review of all possible vaccine options for all possible diseases. Rather, we will explain modern approaches to vaccine development and illustrate these approaches with representative examples. The last section covers the pharmaceutical aspects of vaccines.

16.8.4.1 Recombinant Vaccine

For vaccination generally antibodies producing organisms are inoculated with either killed or attenuated infectious agents including virus or bacteria. Due to certain concern with such kind of vaccine preparation including culturing and handling of the infectious agents at large scale and inappropriate attenuation that can result in spreading of the diseases, recombinant DNA technology is considered more reliable and safer for development of vaccines. Through the tools of this technology, it is possible to delete the pathogenic genes permanently from a disease-causing microorganism. In this way a live microorganism with all features for provoking immunity can be used as a vaccine without fear of reverting to the virulent form. Similarly, genes that are encoding for major epitopes of certain pathogens can be cloned in live microorganism (non-pathogenic) for vaccine development. Genes for antigenic determinants can be cloned in expression vectors for production, purified and then used as a vaccine. In some cases plasmid containing genes for epitopes can be delivered directly to the recipient and in this way expression of the target gene will result in exposure of the antigenic determinant to the immune system.

16.8.4.2 COVID-19 Vaccine

The COVID-19 vaccine is a vaccine designed to provide acquired immunity against severe acute respiratory syndrome coronavirus 2 (SARS-CoV coronavirus 2), the virus that causes coronavirus disease 2019 (COVID-19). Prior to the COVID-19 pandemic, there was an established body of knowledge about the structure and function of coronaviruses that cause diseases such as severe acute respiratory syndrome (SARS) and Middle East respiratory syndrome (MERS), which enabled the acceleration of the development of various vaccine technologies during early 2020. In 10 January 2020, SARS-CoV-2 genetic sequence data was shared through GISAID, and by March 19, the global pharmaceutical industry had announced a major commitment to tackling COVID-19. In phase 3 trials, several COVID-19 vaccines have shown up to 95% efficacy in preventing symptomatic COVID-19 infection. As of April 2021, 16 vaccines have been licensed by at least one national regulatory body for public use: two RNA vaccines (Pfizer-BioNTech and Moderna), seven conventional inactivated vaccines (BBIBP-CorV, CoronaVac, Covaxin, WIBP-CorV, CoviVac, Minhai-Kangtai, and QazVac), five vector vaccines (Sputnik Light, Sputnik V, Oxford-AstraZeneca, Convidecia, and Johnson & Johnson), and two sub-protein vaccines (EpiVacCorona and RBD-Dimer). In all, as of March 2021, there are 308 vaccine candidates in various stages of development, with 73 in clinical research, including 24 in phase I trials, 33 in phase I and II trials, and 16 in phase III development.

References

1. Roberts K, Raff M, Alberts B, Walter P, Lewis J, Johnson A. Molecular biology of the cell: Garland Science; 2002.
2. Clark DP, Pazdernik NJ. Biotechnology: Newnes; 2015.
3. Ratledge C, Kristiansen B. Basic biotechnology: Cambridge University Press; 2001.
4. Brown TA. Genomes 4: Garland science; 2018.
5. Bourgaize D, Jewell TR, Buiser RG. Biotechnology: Demystifying the ConceptsBenjamin/Cummings; 2000.
6. Gad SC. Handbook of pharmaceutical biotechnology: John Wiley & Sons; 2007.
7. Khan FA. Biotechnology fundamentals: CRC Press; 2011.
8. Walsh G. Pharmaceutical biotechnology: concepts and applications: John Wiley & Sons; 2013.
9. Gupta V, Sengupta M, Prakash J, Tripathy BC. An Introduction to Biotechnology. Basic and Applied Aspects of Biotechnology. 2016:1-21.
10. https://www.nature.com/scitable/definition/nonsense-mutation-228/ (accessed 28 March 2022)
11. Sandhu SS. Recombinant DNA technology: IK International Pvt Ltd; 2010.
12. Stein G. Recombinant DNA and cell proliferation: Elsevier; 2012.
13. Davis L. Basic methods in molecular biology: Elsevier; 2012.
14. Glick BR, Patten CL. Molecular biotechnology: principles and applications of recombinant DNA: John Wiley & Sons; 2017.
15. Zhang Z, Tang Y, Yao S, Zhu S, Feng Y. Protein engineering of insulin: Two novel fast-acting insulins [B16Ala] insulin and [B26Ala] insulin. Science in China Series C: Life Sciences. 2003;46(5):474-80.
16. Milne C, Morrison A. Demystifying biotechnology: the PCR game. Teaching Science. 2007;53(1):19-23.
17. Khudyakov YE, Fields HA. Artificial DNA: Methods and applications: CRC Press; 2002.
18. Crommelin DJ, Sindelar RD, Meibohm B. Pharmaceutical biotechnology: fundamentals and applications: Springer Science & Business Media; 2013.
19. Kumar M, DeVaux RS, Herschkowitz JI. Chapter Thirteen – Molecular and Cellular Changes in Breast Cancer and New Roles of lncRNAs in Breast Cancer Initiation and Progression. In: Pruitt K, editor. Progress in Molecular Biology and Translational Science. 144: Academic Press; 2016. p. 563-86.

Index

A

Abrasiveness, 29
Accela-cota and Hi-coater systems, 113
Acceptance criteria, 163
Adhesion, 231
Aerodynamic diameter, 131, 133
Aerodynamic particle size distribution (APSD)
    apparatuses and techniques, 131
    cascade impactor, 131
    deposition pattern, 132
    impact plates, 134
    NGI, 133
    pharmacopeias, 134
    pre-separator, 132
    quality control apparatus, 134
    stages, 134
Aerosol generation, 135
Agitators, 42
Airborne contamination control
    ACR, 194
    clean/aseptic environment, 194
    filtration levels, 194
    filtration system, 194
    LAF/unidirectional airflow, 195, 196
Air changes rate (ACR), 194
Air flow resistance, 138, 139
Air jet mixer, 19
Aluminum, 58
Aluminum tubes, 65
Alveolar macrophages, 127
Alveolar type II cells secret surfactant, 126
Alveoli, 124
Amphiphilic polymers (surfactants), 183
Amphoteric surfactants, 73, 74
Anchor agitator, 18

Andreasen apparatus, 32
Andreasen method, 32
Angiogenic factors, 238
Anionic surfactants, 72, 73
Anticodon, 265
Antimicrobial agents, 71
Antimicrobial preservatives, 159
Antimicrobial substances, 127
Antioxidants, 71, 89, 159
Antisolvent precipitation, 242
Apparent viscosity, 92
Aptamers, 278, 279
APV Gaulin homogenizer, 41
Aqueous-based vehicles, 182, 183
Aqueous phase, 71
Aseptically prepared products, 198
Asymmetric TPX membrane method, 168
Attrition, 28
Auxins, 272

B

Backing laminates, 166
Bacteriophage, 266
Baffles, 18
Bag filter assembly, 52
Ball mill, 35, 36
Benzyl benzoate, 183
Bernoulli's principle, 125
Betaxolol (beta-blocker), 191
Bingham fluids, 87
Biocompatibility, 253
Biocompatible synthetic polymers,
    254, 257
Biocomposite polymer scaffolds, 257

Biodegradable polymer
  biocompatibility, 253
  biopolymers, 253
  drug delivery systems, 257–258
  nanofibrous scaffolds
    for bone tissue regeneration, 257
    for tissue engineering, 256, 257
  natural polymers, 253, 254
  susceptibility, 254
  synthetic (*see* Synthetic biodegradable
    polymers)
Biodegradable polymeric scaffolds, 257
Biological drugs
  cytokines, 280
  hematopoietic growth factors, 279
Biopolymers, 253
Biotechnology-based therapies
  cellular and biomolecular processes, 261
  definition, 261
  DNA, 262, 263
  eukaryotic cells, 262
  gene therapy, 275, 276
  organisms, 262
  prokaryotic cells, 262
Blister packaging, 63
Boiling point of liquid, 12
Bound water, 3
Breath-actuated devices, 136
Bronchodilating agents, 126
Bronchodilating contents, 123
Brownian motion, 128, 129
Buffers, 184

C
Canister, 136
Capsules
  advantages, 116
  dosage form, 114
  hard gelatin capsules (*see* Hard gelatin
    capsules)
  limitations, 116
  liquid/semisolid, 115
  soft gelatin capsules (softgels), 120
Carbopol gel resin, 169
Carboxymethyl cellulose (CMC), 85
Cartridge filter assembly, 52, 53
Cascade impactor, 131, 132
Cationic ion-exchange resins, 226
Cationic surfactants, 73
Cell cultures
  animal, 272
  compounds, 272
  microbial, 271, 272
  plants cells, 272

Centrifugal samplers, 198
Cetrimide, 73
Chemical hazards
  chemicals, 211
  exposure, 211
  hazardous chemicals, 211
  preventive measures, 211
Chemical incompatibility, 85
Chemical penetration enhancers
    function, 160
Chemotherapy, 279
Chewable tablets, 97
Chloramphenicol palmitate suspension, 82
Chlorofluorocarbons (CFCs), 135, 136
Ciliated cells, 126
Clean room requirement, 199
Clean rooms classification
  aseptically prepared products, 198
  cleanliness level, 198
  critical area, 198
  limits, 195, 198, 199
  supporting area, 198
  terminally sterilized products, 199
Coagulation, 83, 84, 89
Coated microneedles, 172
Cold welding, 103
Collapsible tube, 65
Colloid mill, 40, 88
Colony-forming units (CFU), 198
Comminution
  description, 27
  mechanism of size reduction, 28 (*see also*
    Particle size reduction)
  milling/grinding, 27
  particle size, 28 (*see also* Particle size
    analysis)
  particle size distribution, 29
Common health hazards, 204
Common materials, pharmaceutical packaging
  glass (*see* Glass containers)
  metal (*see* Metal containers)
  plastic (*see* Plastic)
Compaction, 102, 103
Complementary DNA (cDNA), 267
Compound annual growth rate (CAGR), 240
Compressed tablets, 97
Compression, 28, 105
Conductivity method, 32
Conjugative plasmid, 266
Consolidation, 103–105
Contact angle, 87
Contact plate sampling, 197
Continuous phase, 67
Controlled flocculation, 83, 89
Convective mixing, 21

Conventional capsules, 117
Conventional coating pan, 112
Conventional film coating, 110
Conveyor dryer, 8, 9
Copolymers, 252
Co-solvent system, 181
Co-solvents, 183
Coulter counter, 33
Coulter principle, 32
COVID-19 vaccine, 282, 283
Cracking, 74, 76
Creaming, 76, 77
Creams, 155
Critical area, 198
Critical moisture content (CMC), 5
Crystal growth, 86, 241
Cunningham slip correction factor (C_s), 127
Cutter mill, 34
Cutting, 28
Cytokines, 280

D
Darcy's model, 46
Deagglomeration, 138, 140
Deflocculated suspension, 91
Degree of flocculation, 91
Delivery systems, 218
    See also Drug delivery systems (DDS)
Demixing, 20, 21
Density, 92
Deposition mechanism, 129
Derjaguin-Landau-Verwey-Overbeek theory
        (DLVO theory), 83
Dermal drug delivery
    creams, 155
    definition, 153
    factors, 155
    ointment, 155
    ointments formulation (see
        Ointment bases)
    physical attributes, semisolid preparations,
        155, 156
    preparation procedures, 161
    semisolids evaluation (see Semisolids
        evaluation test)
Dermis, 152
Dermo-epidermal junction, 152
Dideoxyribonucleotide triphosphate
        (ddNTP), 270
Die cavity, 99, 101
Die filling, 104
Differential pressure, 194
Diffuse layer, 82

Diffusion, 130
Diffusive mixing, 21
Direct compression, 99
Direct dryers, 6
Direct incorporation method, 89, 90, 161
Disc filter assembly, 50
Dispersed particles, 82–84
Dispersed phase, 67
Dispersion medium, 67, 74
Dispersion system, 81
Dissolution testing, 93
Dissolving microneedles, 172
DNA replication
    origin of replication, 264
    transcription, 264, 265
    translation, 265
Dosage ramp, 104
Dosator device, 118
Dria coater, 113, 114
Droplet size, 70
Drug content uniformity, 92
Drug delivery systems (DDS)
    floating (see Floating drug delivery
            systems (FDDs))
    gastro-retentive, 227 (see also Gastro-
            retentive drug delivery systems
            (GRDDs))
    magnetic systems, 232
    mucoadhesive (see Mucoadhesive DSS)
    sustained-release (see Sustained-
            release DDS)
Drug metabolism, 124
Drug nanocrystals, 241
Drug permeation, 189
Drug products, 152
Drug solubility, 181
Drug solutions, 179
Dry granulation, 99, 101, 102
Drying, 100
    adsorbent, 2
    description, 1
    heat transfer and mass transfer equations, 2
    primary and secondary pharmaceutical
            processes, 1
    rate of evaporation, 2
    and vaporization, 1
Drying curves, 4, 5
Drying process
    bound water, 3
    dry-bulb temperature, 4
    end point detection, 5
    equilibrium moisture content, 3
    percent loss on drying (% LOD), 3
    percent moisture content (% MC), 3, 4

Drying process (*cont.*)
    total amount content, 3
    wet-bulb temperature, 4
Dry mixing, 100
Dry powder inhalers (DPI)
    advantage, 137
    design aspects, 137–139
    formulation, 140
    poor drug delivery efficiency, 137
Dry sieving, 100, 101

E
Ejection and Takeoff, 105
Elastic deformation, 102
Electric hazard, 207
    effects of electric shock, 207
    heat, 208–209
    occupational injuries, 207
    preventive measures, 207–208
    sources, 207
Electric hazard electrical injuries, 207
Electrical double layer, 82
Electroporation, 266
Electrostatic repulsive forces, 83
Employee Assistance Program (EAP), 214
Emulsification, theory of, 69
Emulsifier, 69
    complexation of formulation
        components, 74
    improper/insufficient, 74
Emulsifying agent(s), 67, 69
    aqueous phase, 71
    HLB, 71, 72
    hydrophilic emulsifying agents, 71
    oil phase, 71
    surface active agents, 71
    surface-active agents, 72–75
Emulsion instability, 76
    cracking, 74, 76
    creaming, 76, 77
    dispersion medium, 74
    phase inversion, 77
    sedimentation, 76, 77
    types, 74, 76
Emulsion manufacturing, 77
Emulsions
    advantages, 68
    excipients used, 77, 78
    formulation (*see* Formulation of emulsions)
    limitations, 68
    manufacturing, 77
    pharmaceutical, 67

    theory of emulsification, 69
    types, 68
Endonuclease, 266
Enhanced permeation and retention
        (EPR) effect
    active and passive targeted drug
        delivery, 237
    active targeting, 239, 240
    charge, 238
    molecules, 238
    shape, 238
    size, 238
    tumor vasculature, 238
    tumr extravascular environment, 238
Entanglement, 47
Enteric coating, 110
Environmental monitoring systems,
        197, 198
Enzymatic makeup, 124
Epidermis, 152
Epinephrine, 189, 190
Epithelium
    airways, 126
    alveoli, 126
    composition, 126
    goblet cells, 126
    lung, 126
    mucins and proteoglycans, 126
    mucociliary clearance, 127
    smooth muscle cells, 126
    surfactant, 126
Etching, 173
Ethylene vinyl acetate copolymer
        (EVAC), 61, 169
Eye anatomy, 188
    cornea, 188
    iris, 189
    lens, 189
    limbus, 189
    retina, 189
    sclera, 188
    vitreous humor, 189
Eye drops, 190

F
Ferrograd Folic®, 224
Fiber-like particles, 131
Fick's law of diffusion, 16
Fillers, 62
Film coating, 110–112
Filter aids, 47
Filter leaf assembly, 48

Filter media, 47, 48
  nonwoven type, 48
  woven type, 48
Filter press, 49, 50
Filtration
  description, 45
  factors affecting rate of filtration, 46
  filter aids, 47
  filter media, 47
  mechanism of filtration, 46, 47
Filtration devices
  bag filter assembly, 52
  cartridge filter assembly, 52, 53
  disc filter assembly, 50
  filter leaf assembly, 48
  filter press, 49, 50
  rotary drum filter assembly, 51
Filtration mechanism, 46
Fire hazard, 206
  fire safety management, 206
  preventive measures, 206
  secondary manufacturing
        operations, 206
  types, 206
Fitz mill, 38
Floating drug delivery systems (FDDs)
  advantages, 231
  density, 228
  effervescent systems, 230, 231
  limitations, 231
  non-effervescent, 229
Floating test, 135
Flocculated suspension, 91
Flocculating agents, 89
Flocculation, 76, 83, 88, 89
Fluid energy mill, 37
Fluidization of particles, 9
Fluidized bed coaters, 114
Fluidized bed dryer (FBD), 9
  horizontal, 10
  vertical, 10
Fluids, 87
  type, 87
Folate receptor (FR), 239
Formulation development process, 182
Formulation of emulsions
  droplet size, 70
  emulsifying agents (see
        Emulsifying agents)
  preservation, 71
  type, 70
  viscosity, 70, 71
  volume, internal phase, 70

Formulation of suspensions
  chemical incompatibility, 85
  excipients used, 84, 85
  flocculation, 88, 89
  homogenization, 88
  mixing, 88
  particle size, 85, 86
  preservation, 89
  size reduction, 88
  viscosity, 86, 87
  wetting, 87, 88
Freeze dryer, 12
Freeze drying, 12
  advantages, 12
  disadvantages, 13
  uses, 13
Freeze-thaw cycling, 92
Freon effect, 136
Functional film coating, 110
Fusion bonding, 103
Fusion method, 161

G
Gastric emptying, 227, 228
Gastric emptying time (GET), 227
Gastro-retentive drug delivery
        systems (GRDDs)
  FDDs (see Floating drug delivery
        systems (FDDs))
  gastric emptying, 227, 228
  sinking (high-density) systems, 228
  stomach, physiology, 227, 228
  types, 228
Gel formating solutions, 190, 191
Gel formulations, 157
Gene therapy, 275, 276
Gentamicin sulfate cream formulation,
        161, 162
Geometric dilution method, 22
Gibbs free energy, 69
Glass containers
  classification, 59
  composition, 58
  as primary packages, 58
  quality control test
        powered glass test, 59
        thermal shock, 60
        water attack test, 59
Globally harmonized system (GHS), 212
Glucagon-like peptide 1 (GLP-1), 274
Glycosaminoglycans, 238
Glycosylation, 274

Gold nanoparticles (AuNPs), 246, 247
Gravitational deposition, 129
Growth factors (GFs), 280

H
Hammer mill, 35
Hard gelatin capsules
    components, 116
    conventional capsules, 117
    cooling, 117
    cutting, 117
    dipping, 117
    drying, 117
    industrial-scale filling, 116
    joining, 117
    sizes, and filling capacities, 116
    spinning, 117
    stripping, 117
Hazard, 203
    See also Industrial hazards
Heat hazard
    changes, 208
    heat stress, 208
    preventive measures, 208–209
    radiation, 209
    risk of injury, 208
Heating, 2
Heating, ventilation, and air
            conditioning (HVAC)
    airborne contamination control, 194–196
    cleaning equipment, 197
    components, 193
    environmental monitoring systems,
            197, 198
    objective, 193
    personnel contamination control systems,
            196, 197
    pressure differential control, 194
    temperature and humidity control, 193, 194
Heat-shock method, 266
Heavy-duty mixers, 39
Helical impeller, 18
Hematite, (α-Fe$_2$O$_3$), 247
High efficiency particulate air (HEPA),
            195, 196
High-pressure homogenization (HPH), 41, 241
High shear mixer granulator, 24
Hollow microneedles, 172
Hollow microspheres/microballoons, 229
Homogenization, 41, 88, 162
Homogenizers, 41, 42
Horizontal FBD, 10
Horizontal fluidized bed dryer, 11

Hot air, 106
Human health hazards, 203
Humectants, 160
Humidity control, 193
Hydrofluorocarbons (HFC), 135, 136
Hydrophilic and hydrophobic surfaces, 87, 88
Hydrophilic colloids, 224
Hydrophilic emulsifying agents, 71
Hydrophilic-lipophilic balance (HLB), 71, 72
Hydrophilic molecules, 245
Hydrophobic liquid solvents, 157
Hydroxypropyl methylcellulose (HPMC), 229
Hypodermis, 152
Hypromellose, 116

I
Immediate-release film coating, 110
Immersed tube system, 112
Impact action, 28
Impact mill, 35
Impaction, 128
Impeller-type liquid batch mixer, 17, 18
Impingement, 46
Improper/insufficient emulsifier, 74
Indirect dryers, 6
Industrial hazards
    chemical hazards, 211, 212
    electric, 207–208
    mechanical hazards, 211–213
    physical hazards (*see* Physical hazards)
    psychosocial hazard, 213, 214
Industrial noise, 204
Industrial-scale filling of hard gelatin capsules
    liquid/semisolid filling, 119
    powder filling, 118, 119
Industrial-scale manufacturing of suspension
    direct incorporation method, 89, 90
    precipitation method, 90
Inhalation device systems, 143, 144
Inhalation therapy
    advantages, 124
    local and systemic drug administration, 124
    local therapy, 124
    particle engineering, 130–131
    powdered Durata plants, 123
Injectable dosage form, 179
Injectable emulsions preparation, 186
Injectable scaffolds, 257
Injectable solutions preparation, 185
Injectable suspensions preparation
    aqueous vehicle, 185
    sterile in situ crystallization, 186
    sterile powder, 185

Inorganic nanoparticles
   biomedical applications, 246
   gold nanoparticles (AuNPs), 246, 247
   IONPs, 247, 248
Insoluble ocular inserts, 192
Intercellular lipid pathway, 155
Interfacial tension theory, 69
Interferons (IFNs), 280
Interleukins (ILs), 280
Internal phase, 67, 70
Intestinal tract, 124
Ionamin® capsule, 227
Ion-exchange systems (IES), 225, 226
Ion-sensitive gelation, 191
Iron oxide nanoparticles (IONPs), 247, 248

J

Jet nebulizer, 141, 142

L

LAF system advantages, 196
Lamina propria, 126
Laminar airflow (LAF), 195
Laminar flow, 17
Laminar flow hood (LFH), 168
Langevin equation, 128
Laser diffraction (LD), 91
Lecithin, 73, 74
Lighting hazard, 205
Light-scattering techniques, 91
Lipophilic antioxidants, 71
Liposomes, 244–246
Liquid Gaviscon®, 229
Liquid mixing, 19
   See also Mixing of liquids
Liquid/semisolid filling in hard gelatin
       capsules, 119
Lithography techniques, 172
Locally acting drugs, 124
Lubrication, 101
Lymphocytes, 279
Lyophilization (freeze-drying), 191

M

Macromolecules, 124
Macrostructure, respiratory tract
   airways, 125
   alveoli, 124
   Bernoulli's principle, 125
   characteristics, 125
   generational model, 124
   lung surface, 125
   research, 125
   velocity and bifurcations, 125
Madopar HBS®, 229
Maghemite (γ-Fe_2O_3), 247
Magnetic systems, 232
Magnetite (Fe_3O_4), 247
Manual mixing, 22
Mass median aerodynamic diameter
       (MMAD), 132–134
Matrix system, 167
Mechanical hazards, 211–213
Mechanical mixers, 39
Mechanical mixing
   high shear mixer granulator, 23, 24
   horizontal ribbon mixer and sigma blade
       mixer, 23
   tumbling mixers, 23
Mechanism of mixing
   convective mixing, 21
   diffusive mixing, 21
   shear mixing, 21
Mechanism of size reduction, 28
Mesh nebulizer, 142, 143
Mesh number, 31
Messenger RNA (mRNA), 264
Metal containers, 57
   aluminum, 58
   iron, 58
   lead, 58
   non-parenteral pharmaceuticals, 57
   tin, 58
Metering, 104
Microballoons, 230
Microbial spoilage, 76
Microelectrochemical system
       (MEMS), 172
Microinjection, 266
Microorganisms, 76, 89
Microscopy, 30
Middle East respiratory syndrome
       (MERS), 282
Migrating myoelectric cycle (MMC), 227
Milling, 241
Mills
   in pharmaceutical particle size
       reduction, 39
Mixer selection, 16
Mixing
   description, 15
   in manufacturing processes, 16
   mutual mixing, 16
   two-ingredient powders, 20
   V type mixer, 24

Mixing of liquids
air jets, 18, 19
batch type, 17
convection, 17
eddies, 17
features, impeller-type liquid batch
mixer, 17, 18
Fick's law of diffusion, 16
impeller mixer, 17
molecular diffusion, 16
shear force, 17
types, shapes and features of impellers, 18
Mixing of solids
degree of mixing, 20
factors affecting mixing, 21, 22
manual mixing, 22
mechanical mixing (*see*
Mechanical mixing)
mechanism (*see* Mechanism of mixing)
mixing of semisolids, 24
principles of solid mixing, 19, 20
segregation of powder, 20, 21
Mixing process
in pharmaceutical manufacturing, 16
Modified-release (MR) dosage forms, 218
controlled-release, 218 (*see also* Drug
delivery systems (DDS))
gastro-retentive, 218
products, 218
sustained-release, 218
Modified-release film coating, 110
Moist material, 2
Moisture, 2
Molded tablets, 97
Molecular diffusion, 16
Monoclonal antibodies, 281
Monolithic/matrix systems, 223
Monomers, 252, 253, 258
Moving-bed dryers, 8
Mucoadhesive DSS
commonly used polymers, 232
concept, 231
mucous membrane, 231
Mucociliary clearance, 127
Mucus, 127
Multidose inhalers, 138
Multiple compressed tablets, 97
Multiple emulsions, 68
Murine hybridoma technology, 281
Myeloid cell lineage, 279

N
NanoActivator®, 248
Nanocarriers, 238

Nanofibrous scaffolds
for bone tissue regeneration, 257
for tissue engineering, 256–257
Nanonization, 241
Nanopharmaceuticals
nanomedicine, 240
organic nanoparticles (*see* Organic
nanoparticles)
Nanotechnology, 235
Nanovesicles, 243
Natural polymers, 116, 253, 254
Nebulizers
drug solution/suspension, 141
jet nebulizer, 141, 142
mesh, 142, 143
ultrasonic, 142
Neutral mixtures, 16
Newtonian fluids, 86
Next-generation impactor (NGI), 133
Next-generation sequencing techniques, 270
Niosomes, 245
Noise, 204, 205
Nonaqueous vehicles, 183
Non-effervescent FDDS, 229
Nonfunctional film coating, 110
Nonionic surfactants, 74, 75, 183
Non-Newtonian flow, 86
Normal humidity, 193
Novel drug delivery system (NDDS)
controlled drug delivery, 236
delayed-release systems, 236
diseases, 236
gastrointestinal tract, 236
high-pressure homogenization, 242
nanotechnology, 235
passive targeting
microenvironment, 236
nanocarriers, 236
properties, 235
site-specific delivery, 236
targeted drug delivery, 236
Nucleation, 241

O
Occupational health and safety (OHS)
problems, 211
Ocular insert, 191, 192
Odor and taste, 91
Oil in water (O/W) emulsions, 68
Oil phase, 71
Ointment, 155
Ointment bases
additives/alternative vehicles, 157
antioxidants and humectants, 159, 160

carriers, 157
characteristics, 157–159
classes, 157
penetration enhancers, 160
preservatives, 159
Ointments and creams preparation procedures
direct incorporation method, 161
extemporaneous compounding, 161
fusion method, 161
large-scale manufacturing, 161
Oligonucleotides, 277
Ophthalmic suspensions, 191
Organic esters, 157
Organic nanoparticles
drug nanoparticles
application, 242
bottom-up approach, 241
nanoamorphous, 241
nanocrystal, 241
preparation, 241
top-down approach, 241
liposomes, 244–246
nanovesicles, 243
niosomes, 245
polymeric nanoparticles, 242, 244
Oriented wedge theory, 69
Oscillating granulator, 36, 37
Oscillation phases, active setup, 143
Oscillation phases, passive setup, 142
Osmolarity, 184, 185, 189
Osmosis, 184
Osmotic pump system, 225, 226
Ostwald ripening, 86

P
Packaging
common packaging material, 63
common techniques (see Packaging
techniques)
definition, 55
pharmaceutical parameters, 55
primary packaging, 56
properties, 56
secondary packaging, 56, 57
tertiary packaging, 57
wrapping material, 55
Packaging techniques
blister packaging, 63
collapsible tube, 65
strip packaging, 64, 65
Paddle impeller, 18
Parenteral dosage forms
administration routes, 178
advantages, 178

disadvantages, 179
drug delivery, 178
formulation (see Parenteral formulation)
history, 179
manufacturing methods, 185–186
types
emulsions, 180
solutions, 179
suspensions, 180
Parenteral emulsions, 180
Parenteral formulation
buffers, 184
drug properties, 182
excipients, 180
osmolarity adjustment, 184, 185
sterilization, 186
surfactants, 183, 184
types
administered volume, 181
administration routes, 181
onset of action, 181
solubility, 181
vehicle properties
aqueous-based vehicle, 182, 183
co-solvents, 183
nonaqueous vehicle, 183
Parenteral solutions, 179
Parenteral suspensions, 180
Particle close repacking, 102
Particle deposition mechanism
diffusion, 129, 130
drag force, 127
impaction, 128
inhalation therapy, 128
inhaled stream of air, 127
Langevin equation, 128
sedimentation, 129
Stk, 128, 129
Stokes law, 127
Particle engineering
aerodynamic diameter, 131
b-adrenoreceptors, 130
DPI, 131
hygroscopicity, 131
ideal deposition area, 130
size factor, 130
spherical particles, 131
Particle size, 85, 86
Particle size analysis
conductivity method, 32
microscopy, 30
sedimentation, 31, 32
sieving, 31
Particle size distribution, 29
Particle size measurement, 91

Particle size reduction
    factors, size reduction process, 28
        abrasiveness, 29
        feed size, 29
        hardness, 28
        moisture content, 29
        softening with temperature, 29
        stickiness, 29
        structure, 29
    mechanism, 28
    of pharmaceutical dispersions, 38
        agitators, 39
        colloidal mill, 40
        emulsion preparation, 39
        homogenizers, 41, 42
        mechanical mixers, 39
        ultrasonic devices, 42
    of powder
        ball mill, 35, 36
        closed circuit milling, 34
        cutter mill, 34
        Fitz mill, 38
        fluid energy mill, 37, 38
        hammer mill, 35
        mills, 33
        open circuit milling, 34
        oscillating granulator, 36
        trituration, 33
    of semisolids, 42
        agitators, 42
        industrial processing, 42
        triple roller mill, 42, 43
        vacuum emulsifying mixer, 43
Particle's Stokes number (Stk), 128
Passive drug diffusion, 154
Pellegrini pan, 112
Penguin movement, 197
Penicillin G Procaine injectable suspension
        USP sterile vehicle, 185
Percent moisture loss, 170
Percent moisture uptake, 170
Perforated drum coater, 112, 114
Permeation enhancers, 166
Personnel contamination control systems,
        196, 197
pH value, 92
Pharmaceutical dosage, 15
Pharmaceutical emulsion, 67
Pharmaceutical products, 177
Pharmaceutical suspension, 81
Phase inversion, 77
Phosphatidylcholine, 245
Phosphatidylethanolamine, 245
Phosphatidylinositol, 245

Phosphatidylserine, 245
Phospholipids, 245
Photon correlation spectroscopy (PCS), 91
pH-sensitive gelation, 191
Physical hazards, 204
    fire, 206
    lighting, 205
    noise, 204, 205
    pressure, 210
Piston gap homogenizer, 88
Piston homogenizer (PH), 41
Planetary mixer, 24, 25
Plastic
    additives in containers
        antioxidants, 62
        fillers, 62
        lubricants, 62
        plasticizers, 62
        stabilizers, 62
    common polymers used for containers (see
        Polyethylene)
    packaging material, 60
    thermoplastics, 60
    thermosets, 60
    types, 60
Plastic deformation, 102
Plastic fluids, 87
Plasticizers, 62
Pneumatic systems, 10
    spray dryer, 11
Poiseuille's model, 46
Polyacrylamide gel electrophoresis, 270
Polyanhydrides, 256
Polydisperse mix, 141
Polyethylene (PE), 169
    characteristics, 61
    EVAC, 61
    EVAs, 61
    polypropylene, 61
    polystyrene, 61
    polyvinyl chloride, 61
Polyethylene glycol (PEG), 274
Polyethylene polymer, 65
Polymer
    biodegradable (see Biodegradable
        polymer)
    classification, 252
    common biodegradable bonds, 255
    copolymers, 252
    definition, 252
    monomers, 252
    nonbiodegradable, 253
Polymer matrix, 165
Polymerase chain reaction (PCR), 268, 269

Polymeric chemical penetration
enhancers, 160
Polymerization, 252
Polypropylene, 61
Polystyrene, 61
Polyvinyl chloride, 61
Powder deagglomeration, 139
Powder filling in hard gelatin capsules,
118, 119
Precipitation method, 90
Preservation, 71, 89
Pressure gradient, 194
Pressure hazard, 210
Pressure-sensitive adhesive, 166
Pressure-temperature diagram
for water, 13
Pressurized metered dose inhaler (pMDI)
actuation, 136
canister, 135
device structure, 135
formulation, 135, 136
inhalation therapy, 134
metering valve, 135
spacers, 137
Pre-sterilized mixing filling tanks, 185
Primary packaging, 56
Principles of solid mixing, 19, 20
Procan SR®, 224
Product identification, 96
Promoter region, 264
Propellants, 135
Propeller, 18
Protein engineering
computational biology tools, 273
DNA sequence, 273
drug development, 273
muteins, 273
pharmacodynamic profile, 273
post-translational engineering, 274
synthetic biology, 274
wet lab techniques, 273
Protein therapy
antisense, 277, 278
aptamers, 278, 279
EGS, 277, 278
genetic disorders, 276
H. sapiens, 276
molecules, 276
multifactorial genes, 276
oligonucleotides, 277
ribozymes, 277, 278
small-molecule drugs, 276
Proteins, 238
Pseudoplastic/shear-thinning fluids, 86

Psychosocial hazards, 213, 214
Pulmicort Turbuhaler®, 139
Pulmonary drug devices
DPI, 137–140
nebulizers, 141–143
pMDI, 134–137
SMI, 140, 141
Punches, 102

R
Radiations, 209
Recombinant DNA (rDNA)
genetic manipulation, 265, 266
production, 267, 268
sources, 266, 267
transfer, 266
Red blood cells (RBCs), 279
Release liners, 166
Remodeling processes, 127
Repulsion theory, 69
Reservoir, 167
Resistance-induced turbulences, 139
Respimat, 140
Respiratory tract
macrostructure, 124–127
Restriction enzymes, 266
Reticuloendothelial system (RES), 245
Reverse osmosis, 182
Roller compaction, 101
Rotary drum dilter assembly, 51
Rotary drum dryer, 9
Rotary dryers, 8
Rotary tablet press, 105, 106

S
Salt formation, 190
Scale of scrutiny, 20
Seal coating (sealing), 110
Secondary packaging, 56, 57
Sedimentation, 31, 32, 76, 77, 91, 129
Sedimentation volume, 91, 92
Segregation, 20, 21
Selection of dryers, 5
Self-emulsifying drug delivery systems
(SEDDs), 184
Semisolids evaluation test
antioxidants content, 163
apparent viscosity, 164
appearance, 162
dosage units uniformity, 163
microbes presence limits, 163
particle size, 164

Semisolids evaluation test (*cont.*)
  pH, 164
    preservative content, 163
    sterility, 164
    uniformity in containers, 164
    water content, 163
Settling plate sampling, 197
Severe acute respiratory syndrome
    (SARS), 282
Shear mixing, 21
Shear rate, 86
Shear stress, 86
Shear-thickening/dilatant fluid, 86
Sieving, 31
Sifting, 22
Sigma blade mixer, 23
Silverson mixer, 41
Simple emulsion, 68
Single-component polymeric
    scaffolds, 257
Single-dose devices, 137
Single particle optical sensing (SPOS), 91
Single-punch tablet press, 105
Skin
  anatomy, 153
  dermis, 152
  drug delivery, 151, 152
  epidermis, 152
  first line of defense, 152
  functions, 151
  hypodermis, 152
  layers, 152
Slipping plane, 82
Slit-to-agar samplers, 198
Slugging, 101
Soft gelatin capsules (softgels), 120
Soft Mist Inhaler (SMI), 140, 141
Solid microneedles, 171
Solid mixing process, 21
Soluble ocular inserts, 192
Spacers, 137
Spatulation, 22
Specialized dryers
  freeze dryer, 12
Specific DNA techniques
  genome sequencing, 269–271
  hybridization, 269
  PCR technology, 268, 269
Spray coating, 112
Spray dryer, 11, 12
Spray dryer atomizer, 11
Spray drying
  advantages, 11

Stability of suspensions
  appearance, 91
  degree of flocculation, 91
  density, 92
  dissolution testing, 93
  drug content uniformity, 92
  freeze-thaw cycling, 92
  odor and taste, 91
  particle size and shape, 91
  pH value, 92
  sedimentation rate, 91
  sedimentation volume, 91, 92
  viscosity, 92
Stabilizers, 62
Stable suspension, 84
Starch, 116
Static bed dryers, 7
  conveyor dryers, 8
  tray dryer, 7
  truck dryer, 7
  tunnel dryer, 7, 8
Stationary layer, 82
Sterile in situ crystallization, 186
Sterile manufacturing, environmental
    control
  clean room classification, 198
  key parameters, 192
  microorganisms and dust particles, 192
  objective, 192
Sterile ophthalmic preparations
  bioavailability, 188
  eye anatomy, 188
  topical products (*see* Topical ophthalmic
    products)
Sterility, 164
Sterilization, 186
Stern layer, 82
Stickiness, 29
Stoke's equation, 31
Stokes law, 76, 77
Straining, 46
Stratum corneum, 160
Stress-susceptible drugs, 142
Strip packaging, 64, 65
Sugar coating
  candy industry, 109
  ingredients, 109
  polishing, 110
  seal coating (sealing), 110
  sub-coating, 110
  syrup (smoothing/color) coating, 110
Superparamagnetic iron oxide NPs
    (SPIONs), 247

Supporting area, 198
Surface active agents, 71, 76, 183
    amphoteric surfactants, 73, 74
    anionic surfactants, 72, 73
    cationic surfactants, 73
    nonionic surfactants, 74, 75
Surfactant(s), 69, 81, 87, 126, 183, 184
Suspending agent(s), 81
Suspensions
    advantages, 82
    disadvantages, 82
    dispersed particles, 82–84
    formulation (*see* Formulation of suspensions)
    manufacturing, 89–90
    properties, 82
    stability, 91–93
Sustained-release DDS
    advantages, 219
    biological factor affecting design
            strategy, 220
        absorption property of drugs, 221
        half-life of drug, 220, 221
        metabolism of drug, 221
    components, 223
    conventional peroral dosage forms, 218
    formulation, 223
    hydrophilic polymeric matrix, 223, 224
    IES, 225, 226
    insoluble polymeric matrix, 224
    limitations, 220
    monolithic/matrix systems, 223
    osmotic pump systems, 225
    physicochemical factor affecting design
            strategy, 220
        dose size, 221
        ionization, 222
        partition coefficient, 222
        solubility and permeability, 222
    reservoir drug delivery systems, 224, 225
    types, 223
Synthetic biodegradable polymers, 254–256
    biocompatible synthetic polymers, 254
    cross-linked polyesters, 256
    polyanhydrides, 256
    polyesters
        PCL, 256
        PGA, 255
        PLA, 255
        PLGA, 255
Syrup (smoothing/color) coating, 110

T
Tablet coating
    defects, 114, 115

equipment
    conventional coating pan, 112
    fluidized bed coaters, 114
    perforated drum coater, 112, 114
film coating, 110–112
friability, 106
gold and silver, 106
hot air, 106
polymer, 106
purposes, 106, 109
in rotating pan, 106
sugar coating, 109, 110
Tablet compression process
    compaction, 102, 103
    consolidation, 103–105
    direct compression, 99
    dry granulation, 101, 102
    rotary tablet press, 105, 106
    wet granulation, 99–101
Tablet formulation
    excipients and role, 96, 98–99
Tablet press, 104
Tablets
    advantages, 96
    compression process (*see* Tablet
            compression process)
    defects, 105, 107–109
    dosage form, 95
    limitations, 96
    physical and chemical attributes, 96
    types, 96, 97
Takeoff blade, 105
Tamping device, 118, 119
Temperature, 75, 193
Terminally sterilized products, 198, 199
Tertiary packaging, 57
Testosterone, 186
The immersion sword system, 112
Thermally induced phase separation
            (TIPS), 257
Thermoplastics, 60
Thermosensitive gelation, 191
Thermosets, 60
Through Skin Appendages, 155
Tin, 58
Topical ophthalmic products
    classes
        eye drops, 190
        gel formation, 190, 191
        solutions, 190
Topical ophthalmic products (*cont.*)
    formulations considerations, 189, 190
    ocular insert, 191, 192
    reconstitution powders, 191
    suspensions, 191

Topically applied products, 154
Toxic mechanism, 131
TPX asymmetric membrane, 168
Transappendageal route, 155
Transcellular pathway, 155
Transdermal drug delivery (TDD)
  definition, 165
  skin patches (*see* Transdermal patches)
  systemic circulation, 153
  systemic effect, 165
Transdermal medications, 152
Transdermal microneedles
  easy penetration, 171
  insoluble metal alloys, 171
  manufacturing methods, 172
  measurements, 171
  types
    coated, 172
    dissolving, 172
    hollow, 172
    solid, 171
Transdermal patches
  advantages, 167
  components, 165
  manufacturing methods, 168
  quality control parameters
    drug content uniformity, 170
    folding endurance, 170
    percent moisture uptake, 170
    physical appearance, 169
    thickness, 169
    weight variation, 169
  scopolamine, 165
  therapeutic applications, 171
  types
    multilayer drug in adhesive, 167
    single-layer drug in adhesive, 166
Transferrin receptor (TfR), 239
Translation, 265
Tray dryer, 8
Triple roller mill, 41–43
Trituration, 22
Tumbling, 22, 23
Tunnel dryers, 8
Turbine impeller, 18
Types of dryers
  direct dryers, 6
  FBD, 9–10
  indirect dryers, 6
  moving-bed dryers, 8
  pneumatic systems, 10–11
  static bed dryers, 7–8
Typical Fitz mill, 38

U
Ultrasonic nebulizer, 142
Ultrasonic vibration, 42
Ultrasound attenuation (UA), 91
Uniformity of Dosage Units, 163

V
Vaccines, 281
Vacuum emulsifying mixer, 43
Valrelease®, 229
van der Waals forces, 76, 140
Vapor patch, 167
Vegetable oils, 157
Verbal violence, 213
Vertical FBD, 10
Vertical flow room, 195
Vertical fluidized bed dryer, 10
Vibration absorbent material, 204
Violence, 213
Viscoplastic fluids, 87
Viscosity, 70, 71, 86, 87, 92
Viscosity theory, 69

W
Water in oil (W/O) emulsions, 68
Wet granulation, 99
  advantages, 101, 102
  die cavity, 101
  dry mixing, 100
  dry sieving, 100, 101
  limitations, 101, 102
  lubrication, 101
  wet mixing, 100
  wet sifting, 100
Wet mixing, 100
Wet sifting, 100
Wetting, 87, 88

X
Xanthan gum, 191

Y
Yield stress, 87
Yield value, 87

Z
Zeta potential, 82, 83

Printed in the United States
by Baker & Taylor Publisher Services